AF553802

Principles of Physical Geography

Principles of Physical Geography

S. A. Qazi

A.P.H. PUBLISHING CORPORATION
4435-36/7, ANSARI ROAD, DARYAGANJ,
NEW DELHI 110002

Published by
S.B.Nangia
A P H Publishing Corporation
7, Ansari Road, Daryaganj
New Delhi 110002
Ph.: 23274050
E-mail : aphbooks@gmail.com

2025

₹-2595/-

Printed at
Balaji Offset
Navin Shahdara, Delhi 110032

DEDICATED

to my wife Chand Bibi and my children Salma and Navaid, wihout their patience, tolerance and love this book would not have been completed.

Contents

Preface

Physical Geography is the branch of geography which deals with the earth, including land, water and air. As a matter of fact it studies physical features of the earth, its interior, the atmosphere that surrounds it as an envelop and water which is two third of its surface. It brings together the subject matter of many earth sciences to give a general knowledge of the nature of man's natural environment. Hence, it is a combination of mathematical geography, geomorphology, climatology, oceanography and meteorology.

Without physical geography there can be no geography at all. It is the source of the understanding of other branches of the subject. It concerns with rocks, shape and form, configuration, extent of seas and oceans, enveloping atmosphere, without which life on the earth can not exist, physical processes which takes place in atmosphere, the thin vital layer of soil and green vegetation. All these with conjunction, comprise man's physical environment. In other words, it studies man's adaptation to the environment in which he lives and works. Thus, on this basis Ratzel has defined geography as 'the department of knowledge which studies the varied features of the earth's surface as the environment of mankind'.

During the last few decades geography has made rapid strides in terms of its definition, scope, major objectives and methods of study. Today, it is not only a description of regions but an enquiry, a study of causes and effects, an attempt to find out the 'how and why' of facts that go to influence the life of man on this earth. It tries to make analysis of relationship between man and his environment. It depicts the geographical condition under which man lives and works. It is the study of spatial patterns of interactions between physical environment and human activities, so that man can find out the ways to save himself from their adverse responses. It describes how man is conditioned by his environment and how he responds to it. Thus, in this book the emphasis have been given to show the earth as the home of man.

Description, classification and explanation are the objectives of physical geography. Thus, this book gives an appropriate understanding and knowledge of nature and scope of geography, origin of the earth, interior of the earth, earth movements, weathering and erosion, salinity of oceans, waves, tides and currents besides the utilization of wind as a source of energy and hazards of environmental pollution and its remedies.

All the chapters of this book have been dealt with in greater details. The chapter scheme is as follows.

Chapter one is the introductory part of the geography and deals with meaning, history, present and past concepts and its relation with other subjects. The second chapter introduce the origin of the earth. The third deals with earth's interior—fourth with earthquakes, volcanoes and rocks. Chapters 5th to 10th deal with weathering and erosion and development of different landscapes. Chapters 11th to 18th deal with climatology. These chapters introduce the significance of climate in day to day life, composition and structure of atmosphere, temperature, atmospheric pressure, wind, climatic classifications, air masses and fronts. Last chapter deals with oceanography in which marine relief, salinity of the oceans, waves, tides, currents and their effects have been explained in detail. At the end of this book, the glossary of the geographical terms have been given. In this glossary each term has been briefly explained followed by the bibliography and references.

Every aspect has been critically dealt with to make the readers understand the 'how and why' of each phenomenon of the surface of the earth. The maps and diagrams have been designed to bring out the salient facts and are intended to supplement each chapter and to make the subject interesting and pictureque. These diagrams have been compiled from the latest available information. Wherever possible, classifications with careful headings and sub-headings have been used in order to bring some system and ordered relationship into the mass of material with which we are concerned.

This book is the outcome of my experience as a teacher of geography to degree classes. In terms of many Indian University syllabi, it covers a large part of the course in physical geography at undergraduate and post-graduate level. But in general, the book is useful for all readers, teachers and students of geography. I am sure that the book will prove equally useful for common readers and for those appearing in different competitive examinations i.e. IAS, KAS, KPS, NDA, etc.

Any suggestions for its future improvement from all quarters shall receive my best consideration.

Bhadarwah J&K **Author**

Acknowledgement

A teacher is a symbol of spiritual and intellectual enlightenment who illuminates the minds of humanity with the flame of knowledge and enlightenment. The present book is also the result of the sincere, strenuous efforts and guiding force of those great souls who have kindled in me the spirit and love of learning bringing out this book. I owe much to the sincerity, kindness and wisdom of my college teachers Prof. Uttam Singh Bali (Geog.), Prof. B.D. Gupta (Eng.), Prof. Shams-ud-Din (Urdu) and Prof. A.W. Malik (Pol. Sc.).

In preparing this book I have taken a good deal of advantage from the studies of a number of authors and scholars. Their materials were a very helpful guide for me as the manuscript evolved to its final form. I should like to express my thanks to those distinguished authors whose references have been given by me at the end of this book.

A debt of thanks is also owed to Dr. M.N. Kaul, Head, P.G. Deptt. of Geography, University of Jammu for his valuable suggestions from time to time. I owe much to the sincerity of my associate Dr. Javed Iqbal Khatib who skillfully went through and arranged the whole project on the computer and whose immeasurable contributions are really appreciable. My special thanks go to Prof. Basharat and Javed Mughal (English) for their substantial help of proof reading of this edition.

I express my deep sense of gratitude to Dr. M. Iqbal, Dr. Shafqat, Prof. Javed Mohi-ud-Din, Dr. Shuja, Dr. Tariq, Dr. Khurshid Anwar and Prof. S. K. Gupta (Degree College Bhadarwah) for their assistance and affirmative response. Thanks are due to Dr. R.C. Bhagat, Prof. Ab. Rashid, Prof. Zaman, Prof. Mehraj-ud-Din

(Geography) who guided and inspired the author to write this book. I am thankful to library staff of my college particularly to Mrs. Zahida Ji (Chief Librarian) for their active co-operation. At the end I am grateful to the publisher for bringing out this book in time.

Pasri Bus Stand
Bhadarwah - 182222
(J&K). Ph. 244451

Prof. S.A. Qazi
M.A. Geog; M.A. Ur; B. Ed.

1
Introduction

Physical Geography is a broad subject covering many fields. Its most elementary form deals with the location of people, places and things. It is some times called place geography and can be learned by studying maps and globes.

Another important phase of geography is concerned with the distribution of the earth's natural features and human resources. These include mountains, plains, seas, minerals, vegetation, cities and population.

Geography is also concerned with the changing face of the earth. Since the beginning of time there have been violent upheavals of land, erosion of mountain chains, downwarps that formed seas and advancing and retreating glaciers that scoured and changed the land. Man has built cities, dammed rivers, cleared forests and drained swamps.

Its study helps in the assessment of agricultural productivity, regional planning and economic rehabilitation as well as in the preparation and execution of projects in relation to economic growth and social welfare of the country. Its utility has been felt in economic research institutes and planning organizations as its study determines the relationship of resources and activities with land and social goals. Geography provides useful basic material for the techno-economic surveys which are undertaken to make an assessment of development potentialities of the resources of different states. These surveys are meant to show the distribution of resources in relation to physical and climatic aspects of regions and to determined areas where improvement could be possible.

Today the knowledge of geography with all its accompanied methods and techniques is increasingly employed in solving many of the national problems all over the world. The problem relating to floods and drought is the greatest of evils. Though floods cannot be prevented yet flood losses can be reduced and damage can be assessed by studying their aftereffects like the intensity of flood, extent of submerged land, the velocity of the current, the erosion of land and deposition of sediments. This study will help to minimise damage done by subsequent floods.

Geography can bring about a closer co-operation and better understanding among the nations of the world. Geographers have been making valuable contributions to the cause of upgrading of developing countries. The regional variations in the physical features have given rise to regional variations in the field of cultural landscape and in the different ways of living and thinking. The knowledge of geography can be of immense value for planning the economy of a country on regional basis.

At present geographers are not satisfied to learn that kiwis are found in New Zealand or that Dead Sea (Israel) is extremely salty. Now, they want to know that why kiwis are found only in New Zealand and that why Dead Sea is more salty (250 per thousand) than any other ocean. This inquisitive thinking lead to the development of causal relationship in geography. Geographers not only started looking for the causes of the things but went further to note the effects. Today the geographers are interested in cause and effect approaches to various geographical phenomena. They not only explain the surface features of the earth but they also investigate, correlate, rationalise and organise the phenomena of the earth. Geography, today has its own particular field and point of view, hence it has been called simultaneously an art, science and philosophy.

GEOGRAPHY - MEANING AND HISTORY

It is a branch of knowledge that deals with the earth's surface as a home of man. The name comes from Greek and Latin words "geo and graphic", meaning "earth and description," respectively. Hence geography means description of earth. Geography is concerned with the distribution of people, places and things on the surface of the earth, with the relationships between man and his natural environment.

In earlier times its aim was to know facts about places and its people so as to get acquainted with the world. So the descriptions concerning the earth and the people were catalogued under geography. These descriptions were brought through narratives of conquerors, explorers, travellers and traders. Some Greek and Arab scholars laid the foundation of geography as a science by studying the shape and size of the earth, latitude and longitude of the earth and the solar system. Hence, in eighteenth century, the geography was purely descriptive in nature.

Geography is as old as mankind. Man had to think of environment, food and shelter from the very beginning of his existence on earth. There existed certain settled states along the river basins of Nile, Tigris and Euphrates, before Christ. Travellers and conquerors were inspired to know above this inhabited earth, its shape and size and its relation with the universe. Among those who took initiative, the Greek played the most important role by establishing trade and commercial relations with other states as early as the 7th century B.C. On their return from their journeys, they related their accounts of their travels and experiences in the form of stories and poems. Thus the poems of Homer and the stories of Odysseus are well known. At that time it was thought by the Egyptians that the earth was oblong in shape and their own country was its centre.

The following were the notable travellers who contributed substantially to the study of geography.

1. Plato contributed a lot towards human geography by explaining the bearing of physical factors on the character of the Greeks.
2. Aristotle explained the eclipses. He also denied the widely held belief of flat world and expressed the idea of a spherical shape.
3. Pythagoras born is Samos in 582 B.C. was a great Egyptian traveller and was first to find out the shape of the earth and its position in the universe. He said that the earth with *other five planets* is revolving round some fire and that earth is rotating on its axis.
4. Herodotus or Halicarnassus (484-425 B.C.) is known as father of geography. We inherit from him the well known epithet, "Egypt is the gift of the Nile".

5. Alexander, the Great, added much to geography through his conquests and it was he who founded the town of Alexandria. He also prepared maps of his routes and the places he conquered, Thus, contributed to cartography.
6. Roman empire contributed much to mathematical geography. Hipoarchus, the Roman astronomer, introduced the idea of latitudes and longitudes. Strabo, the great historian, wrote an encyclopedia of geographical information in 17 volumes. Ptolemy, as cartographer, evolved the science of preparing maps. The circumference of the earth was calculated as 25000 miles and a map of the world was prepared in the Roman Empire by Eratosthenes..
7. The Christian era did not contribute much to geography except some minor additions. Muslim culture, after this, became important particularly in relation to geography. The Arabs' were the first to show their interest in geography because of their special surroundings. Quite a good number of Muslim geographers added to the subject and prepared manuscripts on geography. One Muslim geographer explained the formation of mountains by folding, sculpturing, erosion and weathering. These Muslim travellers went to China, Japan in the east and to Sicily in the west and gave descriptions of their travels in the form of stories and essays, among which the stories of "Arabian Nights", "Sindbad the Sailor" and "Four Darveshes" are very amusing and full of geographical informations.
8. In the early 9th century Caliph-Al-Mamun of Baghdad encouraged geographical and other kindred studies. Mohammed-Bin-Musa, Librarian in Baghdad, compiled a description of the earth. Another prominent figure was Al-Idrisi (1099-1154) an Arab of Spanish birth, who substituted projections by dividing the inhabited world into seven climatic zones between the equatorial line and the Arctic region, each of which was further divided into eleven equal parts by perpendicular lines[1].

9. Explorations during and after the late 1400's, such as those of Dias, Columbus, Vas-ko-da-Gama and Magellan, proved the world was round and unshared in an age of great discoveries. With it came improved maps and a knowledge of the world never before attainable. One of the great map makers of that time was Gerhardus Mercator, a famous geographer of the 16th century.

Richard Hakluyt, (an Englishman) and Bernhard Varen (a Dutchman) wrote much on the subject matter of geography in 16th and 17th centuries. In the 18th century German philosopher Immanuel Kant was one of the first persons to write on the subject matter of geography in a systematic approach.

As a filed of learning geography thrived and developed many schools of thought, especially in Europe during the 1800's and in the United States around 1900. Among notable geographers were Alexander Von Humboldt, Karal Ritter, Friedrich Ratzel and Albrecht Penck, in Germany; Jean Brunhes and Vidal-de-la-Blache, in France; Sir Halford Makinder, in Scotland; and William M. Davis, in the United States[2].

The trade relations of Christians and Muslims became widespread and established at the end of the 13th century. Hence, the knowledge of an individual about the world was much enlarged as compared to that of Greeks. The existing Arabic literature on geography was translated by the western missionaries and spread over the world. Geography as an independent subject had no place in the curriculum, therefore it was defined as the "description of the earth", by the ancient geographers.

PRESENT CONCEPTS AND TECHNIQUES OF GEOGRAPHY

The ancient Egyptians, Greeks, Romans and others made valuable contributions to Geographical concepts during the 16th, 17th and 18th centuries. More and more geographical concepts developed and geography gradually emerged from a descriptive approach of ancient times to analytical approach of the present times. The discipline of geography at present is overflowing with a number of concepts, which are in the form of its various branches. Geography today consists of a number of well defined generally accepted conceptual elements accompanied by a wide range of methods, techniques and tools capable of presenting and analysing data from different perspectives.

The most widely recognised concept of scientific geography treats the world as essentially an abode of man and solving national and international problems. The perspective of the present day geography is as wide as the earth. The subject matter of the geography is the earth, with rocks, water, universe, people, atmosphere etc. It is because all these have a direct bearing on one another and particular meaning to man, therefore, a need was felt to develop a concept of "applied geography".

Modern geography is now considered to be a separate science requiring a detailed study of the territories of the world. Its instrument of study is the map. The geographers of today are now increasingly concerned with understanding process, patterns and structure and examining geographical data by techniques commonly used in other disciplines. The integration of natural environment with man's social achievements and their expression on the landscape is the field of geographical studies. The modern geography is defined as a unifying science, the raw material it deals with is derived largely from other sciences and studies, it deals with the material in its own way seeking and discovering the inter-relation of phenomena and the interaction between man and phenomena.

The introduction of statistical techniques has proved very useful for carrying out researches in physical, economic, human and regional geography. The land use survey is a technique adopted by geographers for the study of agricultural regions. The statistical technique has also been gainfully applied for determing the indices of crop concentration, distribution, diversification as well as for determination of regions based on these indices. The geographers have evolved a technique of measuring the pressure of population on food resources based on statistical technique, besides the technique of land use survey has been adopted to bring about an improvement in understanding the processes of economic, regional and social development.

A knowledge of geography is helpful for an understanding of current events, foreign relations, why world trade is important.

On the national level geography is an aid to understanding regional differences, which often result in conflicting economic, political and social interests. For example in the United States, the economic interests of the wheat and cattle farmers of the Great Plains are quite different from those of either the industrial population of the East or the Cotton farmers of the South.

The modern concepts of geography will be more clear from the following definitions of geography given by various eminent philosophers.

1. Finch and Trewartha, defined geography as, "Geography is the science of earth surface. It consists of systematic description and interpretation of the distribution pattern and the regional associations of things on the surface of the earth".
2. Ratzel, says, "Geography is the human ecology. Ecology deals with man and environment in which he lives".
3. Ptolemy defines geography in these words that "it is a sublime science which sees reflection of the earth in the heaven".
4. Geography is a pursuit of wisdom in respect of man's adjustment with the environment.
5. The real value of geography lies in the fact that it helps man to live; it helps man to place himself in the world, to learn his true position and what are his duties.
6. Geography enable man to place himself on the world and to know where he stands with regards to his fellows, so that he will neither exaggerate nor diminish his own importance.
7. Geography is the science which treats the relations between the earth and man.
8. The function of geography is to train future citizens to imagine accurately the conditions of the great world stage and so help them to think sanely about political and social problems in the world around.
9. Geography is the subject which deals with the study of earth, its in and out hidden treasures, so that to exploit it for today and tomorrow need.

10. Geography is the subject which describes the earth's surface, its physical features, climate, vegetation, soil, products, people and their distribution.
11. Geography, is a branch of knowledge that deals with the earth's surface as man's home.
12. Ritter, Pesehel and Ratzel, put forth a definition that, "Geography as a whole is regarded as that department of knowledge which studies the varied features of the earth's surface as the environment of mankind".

GEOGRAPHY——HISTORICAL DEVELOPMENT

I. Geography in Ancient Times

Geography in Ancient India: Although geography was not taught as a separate discipline in any University in ancient India, Indian scholars had a very good idea of many geographical and astronomical facts. Our astrologers had a clear picture about the location and moments of various heavenly bodies like planets, moon, sun, starts, etc. There are a large number of scientific treatises on astronomy and astrology like Sury Sidhanta, Panch Sidhanta, Sidhanta Shiromani, Vrihat Samhita etc. At a time when the Western scholars were confident that the earth was flat, Indian scholars know that the earth was round. In 1870 the first experiment was made at the Bedford Level Canal in England to prove the curvature of the earth's surface. On the other hand, Indian astronomers knew more than 2000 years ago that the earth was round. The Hindi equivalent of geography is "Bhugol" which is composed of two words namely "Bhu" meaning "earth" and "Gol" meaning "round", Thus Bhugol means "a study dealing with round earth". In ancient books of learning the word Bhugol appears for the first time in Surya Sidhanta which was written most probably in the 5th century B.C.

The ancient Indian knew about the phenomena of Solar and Lunar eclipses. They had calculated the duration of the revolution of then moon around the earth and of the earth around the sun. They could predict the occurrence of solar and lunar eclipses, correct to a second, hundreds of years ago. In Surya Sidhant, it is mentioned that Rahu and Ketu are two nodel points in the earth's plane of ecliptic where eclipses occur.

Not only the earth and the sun, but they had a fairly good idea about the galaxy also. In Vishnu Puran the shape of the galaxy (group of stats in the sky) is described as like that of a wheel of a chariot. The Pole Star is said to be in the centre of the galaxy - hence named "Dhuru" around which the whole galaxy revolvers like the wheel. There cannot be a better description than the above as the present day scientists agree that the shape of our galaxy is like that of a wrist watch or the wheel of a car.

The Indians, most probably, had an idea about the force of gravity also as it is mentioned in Vishnu Puran that all the stars of the galaxy are tied to the centre of the wheel (Pole Star) by means of an airial cord.

Although their mode of presentation was quite different, the ancient Hindu scholars had calculated the age of the earth very precisely. The Markandeys Puran mentions that 360 days or one year is equivalent to one day and one night of Gods. 360 such earthly years will constitute one year of Gods. 12000 such years of Gods will constitute a Mahayuga. In terms of earthly years, a Mahayuga consists of 4320000 years.

According to Hindu Calender, 1000 Mahayugas (or 4320000000 earthly years) will constitute one day of Brahma. This is the basic cycle in Indian astronomy and is known as Kalpa. Thus 2000 Mahahyugas or 8640000000 earthly years will form one day and one night of Brahma.

In Manusmriti - a Venerable compilation of laws and wisdom- there is a reference that 14 Manus will reign during these 2000 Mahayugas-each period of the reign of a Manu being known as a Manvantra.

According to Manusmiriti, seven Manus have already reigned and seven more are likely to rule. Vishnu Puran mentions that the whole life span of the earth-from its beginning upto its end-is but a day and night of Brahma which is equivalent to 8640,000,000 years of men. Since the earth had already passed through seven Manvantras, it means that the earth had already spent about half of its life and at present the age of the earth is 8640,000,000/ 2=4320,000,000 years. Compare it with the modern scientific estimate and see how correct the ancient Indian Rishis were.

According to the present day geologists the age of our earth is somewhere between 4000,000,000 and 4500,000,000 years.

The Hindus believed in the evolution and dissolution of the Unnetra. Talking of Kalpas, the Bhavishya Puran says, "Excellent sages thousands of millions of Kalpas have passed and as many are to come."

Regarding our world, the ancient Hindus had divided the world into seven Mahadvips namely Jambu, Palaska, Salmali, Kusha, Krauncha, Saka, and Pushkara. The Jambu Dvip is the present continent of Asia of which Bharatvarsha was a part. The Pamir kuvt of Tibet has been identified with the Mount Meru of the Puranic literature.

The whole country from Kashmir to Kanya Kumari and from Burma to Afghanistan was known as Bharatvarsha. The country was divided into nine major divisions, as mentioned in many Purans. These nine major divisions alongwith their present location were as follows:- (1) Indradvipa (Burma) (2) Kasherumat (Bengal and Assam) (3) Tamaravarna (Ceylone) (4) Gabpistiman (Delhi, Haryana and Rajasthan) (5) Naga (Madhya Pradesh and Eastern Maharashtra) (6) Saumya (Uttar Pradesh and Bihar) (7) Gandharva (Afghanistan, Jammu and Kashmir and Northern Punjab) (8) Varuna (Gujarat and North Wester Maharashtra) and (9) Kumardvip (South India).

From the study of Mahabharta, we come to know that at that time the country was divided into 56 principalities, known as Janpadas. The Mahabharta gives a detailed description of all these Janpadas. Not only this, the ancient Hindus were familiar with the topography and the river system of India. The names of Jamuna, Saraswati, Ganga, Gomti, Mahanadi, Godawari, Krishna and Cauvery occur frequently in all religious scriptures.

Our ancient scholars had studied the climatic conditions in different parts of India in detail. Kalidas names six seasons in India. In his Meghdut he gives a very beautiful description of the topography of Central India over which the cloud (Meghdut) neve causing rainfall and bringing cool, refreshing breeze. All the above facts indicate that among ancient Hindus the geographic knowledge of India and of the universe was much more advanced than among the contemporary Arabian, Greek or Roman scholars.

Geography in Ancient Greece: Geography as a specialized field of study in the Western countries, had its beginning in Greece. At the same time, it must be admitted that the Greek scholars borrowed many ideas and facts from the scholars of other countries like Egyptians, Sumerians and Phoenicians. The Egyptians had already developed the methods of measuring land are as and of identifying field boundaries for the purpose of collecting taxes. The Phoenicians (inhabitants of modern Lebanon) were among the earliest explorers and navigators. They brought new ideas about new lands to the people of their motherland, the Sumerians had made some advancement in astronomy. They had collected a large number of facts regarding the positions and movement of heavenly bodies. The year had already been divided into 12 months of 30 days each.

Among the Greek scholars Homer is accepted as the father of geography. The two epics of Homer, namely lliad and Odyssey (perhaps there were two separate Homers who composed these two epics) are great pieces of literature. Whereas lliad is primarily historical, the Odyssey is a geographical account of some known areas of the out side world. Many geographers have tried to identify the places mentioned in the Odyssey. There is a reference in the epic about a land of almost continuous sunshine. Later Odyssey comes to a land of perpetual darkness. These could never have been mere pieces of imagination. Evidently words about the land beyond the arctic circle, where there is six months day and six months night must, have trickled down to Greece.

By about the 8th century B.C., the Greeks started identifying directions by means of wind. Thus, Boreas was the north wind, strong, cool with clear skies. Eurus was the east wind, warm and gentle. Notus was the south wind, wet and violent, and Zephyrus was the west wind, balmy with gale force[3].

Alexander was the first Greek geographer to draw a map of the world to scale. It is true that Sumerians had already drawn "pictorial maps" of some of their cities as early as 2700 BC. But these were not drawn on any particular scale nor there was any true direction. These were merely pictorial sketches. This was_a circular map showing Greece in the centre and other parts of Europe and Asia round it. Alexander is also credited with inventing an instrument, known as Gnomon. With the help of this instrument the varying positions of the sun could be measured by the length and direction of the shadow cast by a vertical pole. This was the fore-runner of the present sundial. The ancient Greeks however thought that the earth was flat like a disc and floating in water.

Herodotus was another great Greek scholar. Basically he is taken as a historian and his great work is a history of Greek struggle against the barbarians; but the work contains a large number of geographical facts. Actually he is credited with the idea that "all history must be treated geographical facts described by him were based upon his personal observation. He had travelled far and wide and had written accounts of many lands. He was the first to describe the formation of delta by river Nile. He reconstructed the ancient shoreline and showed that many former seaports were now far inland.

The two great Greek philosophers, Plato and Aristotle have made great contributions to the development of geographical thought. Plato believed that the earth was round but gave a purely philosophical explanation for that. By that time, the Greek scholars had started thinking that a sphere was the most completely spherical form. Plato agreed that God created earth in a perfect form as the home of man. As God is perfect, he cannot create imperfect objects. Since a circle is the most perfect curve, the earth must be round.

Aristotle, the great scholar was the first to perceive the difference in the habitability of the earth's surface at different latitudes. That habitability was a function of the distance from the equator was a great idea which might have formed the basis of the present natural regions of the world. He also perceived roughly the temperature zone over the earth's surface.

Eratosthener was another great Greek scholar. He is popularly known as the "Father of Geography" as he was the first to coin the word "Geography". He was born in the Greek colony of Cyrene in Libya in about 272 B.C. Eratosthener is most famous for his calculation of the circumference of the earth. He adopted a very ingenious method for it and his estimate was amazingly correct. He improved upon Aristotle's temperature zones by limiting them within certain latitudes. He wrote a book dealing with the various parts of the inhabited earth and also prepared a map of the world.

After his death, Eratosthener was succeeded by Hipparchus as the chief librarian of the museum at Alexandria. Hipparchus was the first to divide a circle into 360 degrees. He was the first to point out that the equator was a Great Circle dividing the earth into two equal halves - the northern hemisphere and the southern hemisphere. Since the earth makes one complete rotation (360 degrees) around its axis once in 24 hours, it means that there is a difference of 4 minutes in the local time between any two longitudes. He is credited with evolving two map projections for showing the curved surface of the globe on a flat sheet of paper.

For the development of geographic thought and the geography of the various lands we owe much to Strabo-a Turkish born geographer who for a major part of his life lived in Greece and had travelled far and wide. He has written the geography of the world in 17 volumes-a monumental piece of work. He has described not only the different lands that he himself had seen, but has also commented upon the writings of earlier geographers. His two books deal with the introductory matter, works of the earlier geographers and his source of information. He had written 8 books upon Europe, 6 upon Asia and 2 upon Africa.

The Roman Geographers: Among the ancient geographers Ptolemy (his real name was Claudius Ptolemaus) holds a place of pride. We do not know much about the life of Ptolemy except that he worked at the library in Alexandria between A.D. 127 and 150. He is the author of a monumental work on astronomy-The Almagist dealing with the position and movements of various heavenly bodies. He believed in the geocentric theory of the universe, i.e. the earth is the centre round which the whole universe revolves. His second book "Guide to Geography" consisting of 8 volumes could be taken as the world's first "Geographical Gazetteer". By that time much information about the distant lands had been obtained by armies, merchants, travellers and scholars. Ptolemy collected all this information and compiled them in the form of tables. He has given not only a descriptive account of various lands but has prepared a new map of the world based on revised data. In this map he used longitudes and latitudes, thus the location of various cities, countries and rivers could be obtained fairly correctly.

II. Geography in the Middle Ages

With the death of Ptolemy the geographic horizons that had been widened by the Greeks began to shrink. For a very long time the development of geography remained in the doldrums.

THE CHRISTIAN GEOGRAPHERS

The Christian missionaries were not interested in the study of geography as such. Their main aim was to reconcile the new graphical ideas with this scriptures, especially the Book of Genesis. Many new ideas of the Greek geographers were at variance with the Holy scriptures. This probably, deterred the Christian scholars to fake up the study of geography seriously. The first medieval Christian writer of geography was Albertus Magnus who wrote a book dealing with the nature of different places. He also discussed the influence of heavenly bodies on the life of mankind and thus combined astrology with environmental determination.

Perhaps the greatest contribution by the Christians to the geographical knowledge was made by the various explorers who sailed to distant lands and travelled through unknown places. Among them the travels of Marco Polo are quite famous. He alongwith his father and uncle started his journey from Venice and reached China through Afghanistan and Central Asia. They remained in China for 17 years and ultimately sailed out of China with a fleet of 14 ships provided by the Mongols Emperor of China. They sailed along the south-eastern coast of Asia, through the Indian Ocean, along the western coast of India and went again to Europe through Persia. Some time later Marco Polo was arrested in Genoa. In the prison he dictated his book of travels to a fellow prisoner. His accounts of China, Japan, Indonesia, Madgaskar and Persia are so detailed and vivid that these could be described as one of the finest works of geographical knowledge.

As time passed with the development of scientific knowledge Christian Geographers began to accept the new astronomical ideas ...eographical facts. Navigation at sea necessitated the ...n of accurate maps showing correct distances and directions. The first mention of the magnetic compass occurs in the writings of Alexander Nekham about 1187. In 1375 appeared the Catalan map of the world which was a great improvement over the previous maps.

THE MUSLIM GEOGRAPHERS

The Muslim geographers had a much more accurate knowledge of the world than Christian scholars. Under the patronage of Khalifa Harun-al-Rahid a project was started to translate works of Greek scholars into Arabic. Under the direction of Khalifa Al-Mamrun, the Arab scholars re-calculated the circumference of the earth. They established a north-south line in the plains of Euphrates. In 921 a Muslim scholar Al-Balkhi collected a host of information regarding climatic conditions in different parts of the world and prepared the world's first climatic atlas—the Kitab-al-Ashkal. Al-Masudi—a famous Muslim scholar had given a good account of the monsoons in India and the South East Africa. Surprisingly enough he has described the process of evaporation and condensation almost very correctly in those days. In 985, Al-Maqdisi—another Arab scholar wrote that there is more of the land in the northern hemisphere than in the southern hemisphere—a fact which came to be known to the European people only in the 19th century after the famous Tetrohydron theory of Lowthian Green in 1875. Al-Maqdisi also divided the world into fourteen climatic regions which was a great advance in the field of geography. In 1030 El-Biruni wrote his famous book "Kitab-al-Hind" — a detailed geography of India. Ibn-Sina, by his studies of the topography in Central Asia propounded the theory that the mountains were constantly being eroded by streams flowing from their tops. The rocks which were hard and resistant to erosion formed the higher peaks of mountains. This idea of the earth's sculpture by river was enunciated by a European geologist, James Hutton, independently some 800 years later.

El-Edrisi was another Muslim geographer who made notable contribution to the development of geography. At his instance King Roger II of Sicily sent many observers to distant lands to make on the spot observations. The observers brought back many new informations. On the basis of these observations Edrisi wrote a

"New geography" in 1154 under the title "Amusement for Him Who Desires to Travel Around the World." He corrected the course of many rivers and the position of many mountains which were wrongly described and depicted by earlier geographers.

One of the most famous Muslim travellers who made great contribution to the geographic knowledge of various lands was Ibn-Batuta who was born at Tangier (North Africa) in 1304. In 1325 he set out for his Huj pilgrimage to Mecca through Egypt. After the Huj he decided to devote himself to travelling. His wanderings took him to Arabia, Red Sea, Ethiopia, and as far south as Mozambique in Africa. He visited Baghdad, Tehran, Bokhara, Samarkand and Kabul and ultimately reached Delhi. For many years he served at the court of Mughal Empire of Delhi. While in India he visited Sri Lanka, Maldiva Islands and Indonesia. Eventually the Mughal Emperor appointed him as his ambassador to China. While in China he toured the country extensively. Returning to Morocco in 1350 he passed through many parts of the Great Sahara and even went to Portugal and Spain. It is estimated that during some thirty years he covered a linear distance of about 75000 miles, which in the fourteenth century was a world record".[1] His travel book is an encyclopedia of the contemporary geography of various lands and people.

THE AGE OF EXPLORATIONS AND DISCOVERIES

Throughout history man has always desired to know something about his neighbouring and distant lands. These explorations widened their mental horizons and added to their geographical knowledge. Fifteenth century could easily be to be the beginning of the age of explorational discoveries, but before that the Vikings from Scandinavia had crossed the Atlantic Ocean and had reached the American mainland. In 874 the Vikings reached Iceland and established a colony there. A little later some of the Vikings from Iceland established another colony in Greenland.

In 1003 a Viking named Karlsafni organised an expedition and sailed to the west with a crew of 160 men and women. They crossed the Atlantic Ocean and landed on the eastern coast of North America. Perhaps they reached as far south as Chesapeake Bay. But the native Red Indians were very hostile and very often fights ensured between the natives and the intruders. Ultimately the Vikings returned home alongwith their rich knowledge of the New World. Among the Norwegians this story is still being sung by the bards under the title of "The Saga of Eric the Red".

THE EUROPEAN EXPLORATIONS

The first initiative for an exploration to distant lands came from Portugal. In 1415 Prince Henry commanded a Portuguese force to the south and captured Ceut a Muslim stronghold south of Jibralter. He then crossed into Sahara and came back with much money and many slaves. By 1455 and 1456 the Guinea coast of Africa was explored.

The great voyage of Vasco-da-Gama took place between 1497 and 1499. By that time the direction and movement of ocean currents were fairly known. Vasco da Gama arrived along the northward flowing Benguela current. He first went westward along the South Equatorial current, then moved southward and then sailed eastward. After crossing the Cape of Good Hope he sailed northward along the eastern coast of South Africa and reached Mozambique. With the help of an Indian Merchant Vasco da Gama sailed across the Indian Ocean and reached Calicut in 23 days. But while he was returning in the month of August he had to face strong south-west Monsoon winds and it took him three months to reach the African Coast. While he finally returned to Lisbon he had sailed 24000 miles in more than two years. Out of an initial strength of 170 men who started with him only 44 survived when he returned home.

Christopher Columbus was a Spainish by birth. He believed that the earth was round and that it was possible to reach India via west. He set out alongwith his fleet towards the west on third of August 1492 sailing through the Atlantic Ocean. After staying at Canary island the party sailed through an unexplored vast ocean. For three months the ships continued their westward journey. On 12th Oct. the fleet reached an island. Columbus was the first to set his foot on this unknown land. Columbus named this land as San Salvador although the native people called it Guanhani. Now-a-days it forms a part of Baluna Islands. Columbus thought that he had reached some island near India and hence he named them as

West Indies although these were close to the continent of North America. The native inhabitants of these islands were red-skinned hunters. On the basis of the same false assumption Columbus called them "Red Indians". The main crop of these island was corn and Columbus gave it the name of "Indian Corn". After visiting several islands in the Carribean sea, Columbus reached Cuba and then Haitti ultimately he returned to Spain on 15th March 1493. Columbus brought with him huge amounts of gold, corn, cotton, many items made from engraved woods, some new kinds of birds and animals as well as few Indians and presented them to the queen of Spain. Encouraged by this success, Columbus made three more explorations, discovered many new islands, landed on the coast of Central America and brought back huge amounts of gold and a large number of Red Indians as slaves to Spain.

The failure of Columbus to reach India by sailing to the west provoked many a youngmen to take up the challenge. By that time the reputation of the East Indies Islands as the suppliers of spices had reached European people. It was Magellan-a Portuguese explorer-who was the first to reach the East Indies Islands by sailing west. Magellan set out on his voyage in 1519 with a fleet of five ships, all in poor condition. After sailing through the Atlantic Ocean, he reached the coast of Brazil. While sailing southward along the eastern coast of South America he entered into river bay to check if it could provide a passage to the west. The fleet passed through the strait of Magellan in October 21, 1520. Coming out of the strait he sailed in the north westerly course and reached island of Guan in ninety eight days. His men were sick with scurvy, they were running short of food and water so much so that they had to eat rats and sawdust. At Guan they replenished their supplies and sailed off again. They reached Phillipines in April 7, 1521. Unfortunately Megallan was killed in a fight with the natives at Phillipines in April 27, 1521. Not only this out of his original five ships, only one survived the remaining four were destroyed on the way. This ship was now commanded by Juan Sebastian who sailed through the Indian Ocean and around the cape of Good Hope went northward through Atlantic Ocean. Finally he reached Spain in July 30, 1522. The cost of the expedition was more than compensated by the huge amounts of spices brought by them.

A little more than 200 years later another courageous sailor Captain James Cook started on his expedition of the Pacific Ocean in 1768. Perhaps this was the first expedition which was undertaken sobly for the purpose of collecting geographical data and for preparing a correct map of the Pacific Ocean. Cook's ship named "Endeavour" reached upto Newzealand and southern Australia.

His second expedition started in 1772 with two ships namely "Discoverer" and "Adventure". The main purpose of this expedition was to discover the continent of Antarctica around South Pole. Captain Cook failed to reach Antarctica but discovered many new islands and prepared new maps of the area.

Cook's third and final expedition started in 1776. With his two ships he reached Cape of Good Hope. From here he entered into the Indian Ocean. After crossing Indian Ocean he entered the Pacific Ocean. In the Pacific he went straight towards north and through the Bearing strait entered into the Arctic Ocean around North Pole. From here he returned to the Hawaiein Islands, where he was killed in a fight with the native people. It is pertinent to note that Captain Cook never reached North Pole. The first person to reach the North Pole was Robert Peary.

III. Geography in Modern Times

Much Pioneering work in making correct large scale maps had already been done by Carrinis in France and by Nicolas Cruquius in Netherlands. In 1817 Adolf Stieler-a German Cartographer published the first sheets of "Stieler Handatlas". It contained 75 maps of the world. Another geographer who contributed to the development of cartography was German scholar Heinrich Berghaus who established a school of cartography at Potsdam near Berlin in 1839.

THE GERMAN GEOGRAPHERS

The Berlin University was set up in 1809, but it was only in 1870 that geography as a separate discipline was introducted in the curriculum of the University when Carl Ritter was appointed the first professor of Geography. Before that Carl Ritter and Alexander Van Humboldt has done much pioneering work in geography. The works of the scholars were available as a base for new geographers. Humboldt had attempted to discuss the distribution of average temperatures over the surface of the earth and had also analysed the influence of temperature on the lives of men as well as on natural vegetation. Carl Ritter had done a lot of research in collecting geographical facts from various parts of the earth and in compiling it in a book form "Erdkunde". He prepared several maps based on the latest information.

Ferdinand van Richthofen was famous German geographer who worked in many Universities. He made a detailed study of the geology of China and located many areas from where coal could be produced. It was he who first of all discovered the Loess region in the north-western part of China. His studies of China were published in five volumes between 1877 to 1912. In his works he had also discussed the nature and scope of geography. He was appointed as a professor of geography in Berlin University in 1886. Being a close field observer, he wrote many articles on the processes which go to shape the surface of the earth.

Whereas Richthofen laid the foundation of the systematic study of the physical features of the earth, it was Friedrich Ratzel who provided guidelines for a systematic study of Human Geography which he called Anthropogeography. In 1874-75 he went to U.S.A. and Mexico and studied the living of minority groups there like Indians, Africans and Chinese. In 1880 he became professor at Munich University and in 1882 he published his first volume of Anthropogeography. Ratzel studied the human societies in relation to their physical environments. He became a professor of geography at Leipzig University in 1886. He died thus in 1904.

Alfred Hettner was another noted German geographer who started a research magazine 'Geographische Zeitschrift' in 1895. He contributed richly to the problem of methodology in geographic study and ultimately published all his articles in the form of a book in 1927. It was he who defined geography as a "Chorological Science".

THE AMERICAN GEOGRAPHERS

In U.S.A. the geographical concepts developed with the writing of George Perkins Marsh. Basically he was a politician but had wide ranging interests. Besides English he could read and write in twenty other languages. He had studies the works of Humboldt, Ritter and Guyot. Although Marsh agreed that natural environments have a great influence on the lives of men, yet he asserted that man need not be slave of nature. He paid attention to man's influence on nature and to the modifications of the organic and the inorganic parts of the human habitat that have resulted from the action of mankind. He collected many examples of the constructive and destructive actions of man on land and other natural resources.

Mathew Fontaine Maury was another scholar who made notable contribution to the development of geography in U.S.A. Unfortunately he had no regular training and as scientist he was self-taught. His special field of study was climatology. Being a director in a naval department he had specific interest in the study of winds and currents. He derived new instruments to obtain the depths of oceans and prepared the first map of the floor of the Atlantic Ocean. His researches in the direction and velocities of winds were of practical value to the trading ships. Maury prepared a diagram of the planetary winds blowing between certain latitudes over the surface of the earth. In those days it was a great contribution to geographic knowledge.

During 1866 and 1879 many expeditions were organized to survey accurately various parts of U.S.A. Among the leaders of the surveying parties Ferdinand V. Hayden; Clarence King; George M. Wheeler and John Wesley Powell are more famous.

Perhaps the greatest contribution to the development of geographic knowledge in U.S.A. was made by William Marris Davis. He is credited with the idea of "Cycle of Erosion" which he called "geographical Cycle". According to him "landscape is a function of structure, process and stage." His major contribution is in the field of geomorphology. It was he who described the three stage in the evolution of topography namely youth, maturity and old age. As a member of the committee appointed by the National Educational Association to consider the courses of study in geography in schools and colleges, he was largely responsible for the drafting of the report of the committee.

Mark Jefferson—a student of Davis deserves our special praises for his contributions in the field of geography. He was professor of geography at the Michigan State Normal College fiom 1901 to 1939. During his thirty eight years of professorship he had a hand in training a large number of teachers of geography. His main emphasis was a man rather than the earth. He pleaded before the Committee Education that the focus of geographic study should be "man on the earth" and not "the earth and Man".[1]

One of the brilliant students of Jefferson was Isaiah Bowman. After teaching for some time at the Yale Forestary School, he went to South America for a field study of the Andes. He spent many months in Peru, Bolivia and Chile studying the structure and physiography of the Andes for his doctoral thesis. He was member of the expedition who re-discovered the lost ancient Inca futures of Mochu Picchu.

Ellsurath Huntington was an associate of Bowman at Vale University. His main field of study was the affect of climate on man's life. He is credited with the idea that our climate is going drier since the Pleistocene glovial period. He had toured the Asian countries extensively. Correlating the historical fact with geography he maintained that the out pouring of the nomadic Mongols from Central Asia leading to their conquest of Eastern Europe, India and China was due to the dessieatun of Central Asian lands. His famous book "Principles of Human Geography" is still considered as a standard treat book for colleges and universities.

J. Russell Smith - originally an economist has also made great contribution in the field of Economic Geography. The first edition of his famous book "Industrial and Commercial Geography" appeared in 1913. Since then the book has run into numerous editions. In addition to this he published 29 other books. His contribution in the field of Economic Geography is rather unparalleled.

During recent years the most notable contribution towards the nature of geography and its methodology has been by Richard Hartshorne. His monumental work "Nature of Geography" stands as a landmark in the history of geographic thought. Hartshorne's book has been unduly acclaimed as an authoritative work. The widespread interest that the book around invited certain criticism also. Hartshorne has tried to answer these criticisms in his new book, "Perspective in the Nature of Geography".

THE FRENCH GEOGRAPHERS

Vidal De La Blache is said to be the architect of modern geography in France. He got his doctorate in geography in 1872 and was appointed professor of geography in 1898. In 1899 he presented his ideas regarding the scope and purpose of geography. He had opposed the idea of environmental determination. Instead he believed in "Possiblism". His book on geography is a monumental work of those days.

Jean Brunks was a celebrated student of La Blache. He not only spread the ideas of his teacher but added his own to the field of geography. He had written a very standard work on "Human Geography." Among the present day other geographers of France the names of Phillips Linchemel and Jean Gothmann are notable.

THE BRITISH GEOGRAPHERS

H.J. Mackinder was the first geographer to be appointed as reader in Geography at Oxford in 1887. He had his earlier training in history, but he believed that history without geography was mere narrative. Since every event had occurred at a particular time and at a particular place, history and geography which deal with time and place respectively should never be separated.

Whereas Mackinder has been accepted as the "Founder" of British Geography, L. Dudley Stamp has been acclaimed as the "Father" of British Geography. His contribution to the study of geography has been unparalleled. He had travelled widely throughout the world. His interest in the underdeveloped countries of Asia, Africa and Latin America has evoked world wide acclaim. He has written a number of books catering to the needs of students at various levels of education. But perhaps the greatest contribution of L. Dudley Stamp was his idea of "The Land-Use Survey of Great Britain". With meager resources and a few trained personnel he set upon his herculean task of preparing the land-use map of Great Britain. About 22000 school and college students who volunteered for the job were given training by him. The work was started in the summer of 1931 and mapping was completed by the end of 1935. These maps showed the various categories in bright colour representing the various uses to which the land of Britain is put to. These maps were prepared on the scale of 1" to a mile. Very soon the usefulness of these maps was recognised by the British Government and the Ministry of Agriculture sanctioned funds for the publication of these maps. His famous book "Land of Britain its use and misuse" has been acclaimed by all geographers as a most standard work.

The Indian Geographers

The introduction of geography as a separate discipline of study in Indian Universities is of recent origins. Although History and Geography, as two papers of the same subject, were taught in most of the teacher-training colleges there was hardly any graduate teacher in geography. It was the prerogative of the teachers in history to teach geography also.

For the first time, geography was introduced as a separate subject at the graduate level at Aligarh Muslim University in 1936. It was followed by a graduate course at the Calcutta University in 1941 and at Madras University in 1948. Now-a-days geography is being taught at the graduate and Post-graduate levels in almost all the Universities of India. Among the noted Indian Geographers who have made substantial contribution towards the development of geography in India, are Prof. S.P. Chatterjee of Calcutta University, Prof. George Kuriyan of Madras University, Dr. R.L. Singh of Banaras Hindu University, Dr. Muzzaffar Ali of Saugar University, Dr. Mohd. Shaffi of Aligarh Muslim University, Aligarh and Prof. Munis Raza of Jawahar Lal Nerhu University, Delhi, All these persons are pioneers in the field of Indian Geography.

Prof. S.P. Chatterjee deserves special mention for setting up the National Atlas Organisation of India. This organisation published a very detailed atlas of India on a fairly large scale sometime in 1953. Unfortunately it has not been revised so far. The Organisation has also published very beautiful and very useful maps, covering the whole of India on the scale of 1:1000000. These maps have been indeed in three series namely - (a) Population Series, (b) Physiography Series, and (c) Transport and Tourism Series.

In the field of geographic research perhaps Dr. R. L. Singh must find a place of pride. Although his dissertation was on Urban Geography, there is hardly any branch of geography in which he has not published a paper. His main interest lies in Urban

Geography, Settlement Geography and Geomorphology. He has written many books for the College and University students. His book "India, a Regional Study" is indeed a monumental work.

Prof. George Kuriyan-an old dedicated geographer has published numerous research paper on diverse topics. He also edited a good atlas for school children.

Dr. Muzaffar Ali is known more for his book "Geography of Puranas" than for his research papers. It is a tribute to a Muslim geographer writing such an authentic account of the Hindu scriptures.

Prof. Munis Raza's interest lies chiefly in Regional Planning. He was chairman of many committees, connected with the department of the subject of geography and was engaged in guiding research in Regional Planning.

GEOGRAPHY AND OTHER SUBJECTS

Geography and History

Geography deals with the space factor or places and history, with the time factor or periods, in the story of man. Geography describers the stage on which human life is enacted while history describes the drama of human life. Thus there exists close relationship between these two subjects. No history can be complete without some reference to space. Similarly no geographical account can be intelligible without reference to development in time. Thus both history and geography are concerned with the interplay of human and physical factors.

Then it is through the study of geography that we can know how the physical features of a country have influenced and can influence the course of history in that country. The teacher of history like a geography teacher, must use maps, charts and diagrams etc. to show the extent of empires, political boundaries, routes of invaders, strategic position of certain important towns and plans of battle-fields etc. These material aids are common to both these subjects. Thus geography is very helpful in explaining a number of historical events.

Geography and Political Science

Political science is the subject which studies the foundations of the State and the principles of government. Sometimes political administration differs from country to country. This difference is due to geographical conditions. In different countries traditions, political principles and certain other things are also guided by geographical factor. Similarly physical conditions greatly influence the character, the national life of the people and other political institutions. Rousseau tried to establish relationship between the climatic conditions and the forms of government when he said that warm climates are conducive to despotism, cold climates to barbarism and moderate climates to a good polity. In his book "History of Civilization", Buckle has asserted that actions of individuals and societies are influenced by the physical environments, particularly climate, food, soil and the general aspects of nature.

Then in the present day world, world citizenship and world government have become the ideals of Political Science and practical politics. Geography also aims at establishing world citizenship by explaining the common geographical features of different countries. Hence there is close relationship between these two subjects.

Geography and Economic

In present age of science and technology, human activities are influenced and controlled more by economic forces known by natural forces. It is, therefore, that in geography more attention is now paid to economic conditions, which modify the influence of natural environment. For this purpose, Economic Geography, as an important branch of geography, is given more stress than other branches. In Economic Geography, various principles are described, on the basis of which it is possible to study the economic aspects of the geographical factors. In other words, economic geography is the study of man's economic activities, as controlled or influenced by physical environment. Agriculture, minerals, forests, industries and trade are all influenced by geographical factors. So economics, which is called the science of wealth, cannot be properly studied without the knowledge of geography. It is the geographical factors which govern the wealth of a particular country or nation. Thus geography provides background to economics. So geography and economics are correlated with each other.

Geography and Language and Literature

Since ages, literature has been very much influenced by geographical conditions. It is full of natural phenomena in the form of poetry, drama, epics and stories etc. the Rig Veda and ancient Sanskrit literature are full of beautiful descriptions of rivers, rain, wind, clouds, mountains, sun, moon, earth and sky etc. Those ancient poets and writers could not have described nature so beautifully and successfully, if they had no knowledge of geographical facts and factors. Thus geography provides rich subject matter to literature.

The language and literature are very helpful for the study of geography. Reading and writing are very helpful tools in geography because printed material and written expression are of information and instruction. It is, therefore, that a child is taught reading as a part of geography instruction. He is also given training in oral and written expression. It is through this medium that we can correctly evaluate the real grasp of the subject matter, on the part of our pupils. In teaching geography, we provide many opportunities to our pupils for discussing, speaking, debating, paper reading and narrating their observations and experiences in black and white. Thus language facilitates the study of geography, especially in secondary classes. Geography also adds many new words and terms to the, vocabulary of language and literature. In this way, geography and language and literature are inter-related.

Geography and Art and Craft

Practical geography provides a vast relationship with art and craft. Drawing of diagrams, outlines, graphs, charts and maps etc. is an important part of geography instruction. Through these valuable aids, we can make the teaching of geography simple, effective and interesting. It is possible only through art activities. The craft helps in making models of lands, dams, canals, wind vane, sun dial and means of communication etc. Thus art and craft are closely related with geography. Models may be prepared of clay and plaster of Paris for depicting various geographical phenomena.

These models may depict rotation and revolution of earth, solar and lunar eclipses, volcanoes glaciers and the like. The teacher of geography must possess illustrative talent in him. Then and only then, he can arouse the interest of his pupils in preparing and using various visual aids for their geography lessons. Thus art and craft provide a base to the teaching of geography and make the subject delightful and interesting.

On its part, geography also provides art and craft with interesting subject material. Various facts of the subject matter of geography may inspire students of art and craft to draw fine pictures of mountains, rivers, dams, lakes, sun-rise, sun-set, water-falls, crops, flowers, harvest-scenes, forests and the like. All this establishes inter-relationship between these two subjects.

Geography and Mathematics

Some important problems in geography cannot be solved without reference to mathematics. Geography involves calculations, survey, meansurements and the use of graphs. All this constitutes training in the use of elements of mathematics in geography. Mathematics gives a language of figures to geographical data, especially connected with population, agriculture produce, exports and imports, mineral production and distribution of rainfall etc. Surveying is almost based on mathematical foundation. Then for measuring the distance, width or depth of rivers, oceans or lakes etc., or the height and distance of certain stars etc., a geographer has to get help from mathematics. Mathematics is also very helpful in giving an idea of latitudes and longitudes as well as in finding out time at different places of the world. This clearly indicates that geography and mathematics are deeply related.

Geography and Natural Science

As we have already noted, geography is a link subject between social sciences and natural sciences. Many facts of geography are included in the subject matter of natural sciences like Physics, Chemistry, Botany, Zoology, Geology, Astronomy and Agriculture etc. Then cause and effect relationship which is so emphasised in the field of geography, is a gift of science. The natural or physical geography which studies climate, pressure and velocity of wind, soil, rocks, currents, weather, mineral wealth, rotation and revolution of the earth, phenomena of earth-quakes and volcanoes, the distinction of flora and fauna, is very closely related to physical chemistry, botany, zoology and geology etc. Then the study of geography involves the handling of various apparatuses, such as Barometer, Thermometer and Rain Gauge etc. These apparatuses cannot be handled and used properly without some knowledge of the principles of Physics. Thus geography gets all the knowledge from natural sciences, which it can effectively utilize in explaining various natural phenomena.

Nature and Scope of Geography

Before trying to acquire geographical knowledge, one must be clear about what he is expected to know. What are the limits and boundaries of his subject. Because unless one knows his field of study, he is likely to transgress into the domain of other sciences and allows himself to be ridiculed. Thus every geographer must know the nature and scope of his subject. As an eminent American geographer Richard Hart Shorne has noted, "Even more important is the need for each individual who proposes to devote his professional life to the field of geography, to have clear picture of the scope and nature of that field[5].

NATURE OF GEOGRAPHY

I. Geography as an Earth Science

The branch of geography dealing with the weather phenomena and relief of the earth is known as Physical Geography on account of which it is also termed as earth science.

Geography has always been and still is essentially a science of mankind. Because its centre of study is man. It studies various physical features and weather phenomena because these exercise a great influence on the life of man, including his health and occupation. His food and clothing will be determined to a large extent by his natural environment. Even his social customs and religious beliefs will be influenced by his physical environments. It is for these reasons that geography studies the earth and atmosphere only to that extent which has a vital bearing on man's living. Geography studies only surface features. It seldom studies the deeper interior parts of the earth, as it is the field of geology. The study of rocks and soils is essential for mineral wealth and agricultural crops respectively for man. From the field of astronomy, geography does study only the rotation and revolution of earth as it is responsible for the formation of days, nights and seasons, which has influence on man's living.

Similarly, geography studies the surface movements of ocean water as these help in trade and fishing. Weather and climate exercise the greatest influence on human life, so its study is essential in geography. Thus we find that geography deals only with the surface of the earth which is the meeting place of its three parts namely lithosphere, hydrosphere and atmosphere, which is the abode of man. It is more clear from this definition "The geography is the study of earth as the Home of Man."[6]

II. Geography as an Environmental Science

"Today geography is recognised as a science which involves a study of the relationship that exist between man and its environments."[7] Thus the study of geography involves the study of all the three types of environment i.e. physical, biological and social environment. Physical environment includes mountains, plains, rivers, temperature, pressure, soils, minerals, size and shape of earth. Biological environment include vegetation and wild animal life and the social environment includes the occupations of people, communication, transport, standards of ethics and morality.

The natural environment play vital role in modifying the life and habits of mankind. This led the earlier geographers to propound the "theory of determinism," according to them man is a slave of nature. The earlier geographers like Huntington and Ward went much further and emphasised that climate effect even the races, physical and mental capabilities and psychologies of the people.

III. The Influence of Climate on Man

It has been accepted by medical science that high temperatures, combined with high humidity make the climate enervative, as it saps the energy of the people. On the other hand dry climate, hot or cold, is said to be good for health. A relative humidity of 50 to 60% is considered ideal for maintaining good health. This is the reason why Bengali and Biharis are thin and lean as compared to those of Punjabi and Kashmiris, who are study and well built.

Climate is the most dominating factor of the environment. Climate directly and indirectly affects the different human activities in his daily life. Climate determines our food, clothing shelter and general mode of life.

The climatic conditions are clearly reflected in the dress of people. The type of clothing varies according to the changes in the weather and climate. In hot humid equatorial regions, man lives half naked. In cold areas, people use light woolen clothes. In Tundra region people put on fur clothes to protect the body against cold.

A Rajasthani protects his neck and head from the beating sun stroke by wearing a long turban. Bengali wears a loose dhoti and Kurta of mulmul. A resident of Kerala moves barefooted in a lungi and a Kashmiri wears woolen clothes to protect himself from intense cold.

Man builds his house to suit the various elements of climate such as temperature, rainfall etc. The roofs of houses at Simla and Kashmir are slanting to drain off the rain water. Flat-roofed houses are built in hot regions. The wooden houses of Japan and Igloos of Tundra (polar regions) reflect the influence of climate on man. In areas of dense forests (equatorial), man-lives on branches of trees.

Climate helps to determine our food. The food habits of the people vary from season to season. In cold areas, people depend on animal food to keep themselves warm. Rice is the staple food in monsoon climate. Dairy products are widely used in cool, temperate climate. In hot and humid lands, people depend on vegetable food and are vegetarians.

The climate influences the food habits of the people, because it influences the agriculture. Rice and fish is the staple food of Bengalis and Biharis, while chapati (Wheat and Maize) is favourite to Punjabis. Cotton grows mainly in Maharashtra, Gujarat and Madhya Pradesh, while Uttar Pradesh is famous for sugarcane. Citrus fruits in Mediterranean region and bananas in the tropical regions reflect the influence of climate.

IV. "Science of Relationship between Nature and Man"

One of the chief tasks of geography is to study the relations between nature or natural environment and man. These relations, sometimes called land-man relations are important because both man and nature play a role in creating the differences and similarities between the various regions and countries of the world.

Topography, landforms and climate have strong influences on the ways man uses the land. For example, landforms helps to determine the location of farming and mining regions and routes of major transportation lines. Some landforms, such as mountains, are barriers to trade, communication and exchange of culture. The relationship between man and such features of the natural environment as plants and animals, soils minerals and water resources are also studied in geography.

The geographers now were no longer satisfied to learn that "Kiwis" are found in Newzealand or that Dead Sea was extremely salty. They now wanted to know that why Dead Sea is more salty than other oceans. This thinking lead to the development of casual relationship in geography. Geographers not only started looking for the causes of things but went further to note the effects. Even today the geographers are interested in cause and effect approaches to various geographical phenomenon.

Today geography not only explains the surface features as they are but it also investigates and correlates the phenomenon of the earth. Today it is descriptive and analytical study. Moreover it has its own particular point of view. Hence, geography has been called simultaneously an art, a science and a geography. It is a developing science, According to Finch and Trewartha, "Geography is the science of earth's surface. It consists of systematic description and interpretation of the distribution pattern and the regional associations of things on the surface of the earth". Ratzel, another geographer says that the "Geography is the human ecology. Ecology deals with man and environment in which he lives". Ptolmy, another geographer says that "Geography is a sublime science which sees reflection of the earth in the heaven." The geography has been mainly divided into two branches i.e. Physical Geography and Human Geography. The former deals the natural aspects while the later is concerned with the man made aspects of geography. Physical geography deals with the study of mountains, rivers, vegetation, soils, atmosphere etc. On the other hand the human geography studies the influence of environment on human life. Man tries to modify his environment and creates man made and cultural features. These features include human settlements, agriculture, means of transport, industries, etc. On the basis of the above assumptions the physical geography is sub-divided into different branches like Geomorphology, Climatology, Hydrology, Pedology, etc. While human geography into Economic Geography, Population Geography, Mathematics Geography, Historical Geography, Political Geography, Cultural Geography, etc.

SCOPE OF GEOGRAPHY

A few decades back the study of geography implied only a catalogue of names of continents and oceans, the depths of seas and bays, the length of rivers and height of mountains. Both the student as well as a teacher of geography felt quite satisfied if he could remember such facts. They never attempted to understand why and how those facts have come into existence. Geography was recognised only as description and interpretation of the regions of the world.

On the contrary geography of today is an enquiry, a study of causes and effects, an attempt to find out the way and how of all those facts and factors that go to influence the life of man on this planet. Modern geography studies earth as the home of man. It attempts to find out the relationship between man and his environment. It depicts the geographical conditions under which man lives and works. In other words, it studies man's adaptation to the environment in which he lives and works.

Geography thus takes a very wide outlook and tries to interpret the action and interaction of all physical factors in relation to the intricate problems of the life of man on the surface of the earth. There lies the utility of studying geography because the discipline of geography brings us into contact with the action of the living man. It brings within its field of study every form of human occupation and every effort directed to harness nature for the progress and development of mankind. Geography is now-a-days considered to be a distinctive discipline able to take its position at par with the physical sciences and humanities in terms of current innovations. The scope of geography has became so vast and complex that a need has arisen for specialization. As a result the subject matter has broken up into a number of branches, such as physiography, climatology, oceanography, human, economic, historical, social, political population, urban, settlement and biogeography.

Geography today covers a vast field and comprises many branches of scholarship in its fold. Like the bee it sucks honey from every flower. Its subject matter consequently lends to and borrows interest from both scientists and students of social sciences, as it includes physical sciences like physics, chemistry, mathematics and astronomy on one hand natural and humanistic studies like botany, zoology, anthropology, sociology and history on the other. Hence Dr. Scott Keltie agreed with the definition of geography put forward by Ritter Peschel and Ratzel that "Geography as a whole is regarded as that department of knowledge which studies the varied features of the earth's surface as the environment of mankind". In 1908, a committee of the International Geographical Congress issued a long definition of the subject, the essence of which is, "Geography as a branch of knowledge has for its object the description of the surface of the earth as evidence of man's relationship with the earth's surface, the home of man".[8]

In other words the scope of geography extends over where, why and how of things found on the earth's surface. The subject has a wide range of its domain which has been sometimes considered too wide by various eminent philosophers and geographers. However, according to Alfred Robinson the scope of geography covers the following areas:

(a) The size, shape and movements of the earth.
(b) The position and distribution of land masses and water bodies.
(c) The structure of rocks and relief of the earth's surface.
(d) The water of the oceans, seas and lakes and their movements.
(e) Distribution of animals and the vegetation cover on the earth.
(f) The atmospheric conditions and the resultant climates.
(g) Human races and the distribution of population.

(h) Activities of man to earn a livelihood.

(i) Human shelter in the form of human settlements and their types.

(j) Society - Human groups, their social and cultural characteristics.

(k) Political organisations and their relationship with human groups.[9]

The geographers think of the causes leading to the difference in the racial, physical and social characteristics of mankind of the different countries of the world. They also study the crops grown and gathered in different parts of the world. This brought the study of weather elements and weather phenomena within the orbit of geography.

Essentially geography is a study of mankind. The home of man is the meeting place of the three parts of our mother earth - i.e. lithospheres, hydrosphere and atmosphere. Thus geography studies all these three parts of the earth. Not only this, the study of various heavenly bodies like the sun, the moon and other planets, their sizes and their movements became the exclusive privilege of geographers.

With the passage of time our knowledge of the various countries and the various features of the earth's surface increased tremendously. There was an explosion in our knowledge about our solar system. This led to the specialization by geographers in different branches. Ultimately astronomy, geology, meteorology and oceanography became separate sciences; but the mother science - geography, still draws heavily upon these specialized sciences.

To sum up, in simple words the scope of geography is the whole world itself, or earth surface with water and atmosphere.

MAIN DIVISIONS OF GEOGRAPHY

Geography can be divided into two categories according to the method of study; the subject can be approached systematically or regionally. Either approach will involve physical geography human geography, or both.

1. Regional Geography

Instead of dealing with a single element regional geography focuses attention on a particular country or region. The goal of regional geography is to view an area in its totality including the natural environment, the people, their economic and cultural activities.

2. Systematic Geography

It keeps the entire world in view while investigating a single element, such as landforms, climate, agriculture and manufacturing. For example, a systematic study of climate includes such things as the world distribution of temperature, precipitation, pressure and winds. It also classifies climates and deals with their distribution through out the world.

The subject matter of systematic geography can be divided into two broad categories i.e. physical geography and human geography. Each has many subdivisions, some of which overlap. Some are closely akin to the physical, biological and social sciences.

I. PHYSICAL GEOGRAPHY

Physical geography is the study of land, water, surface features and the natural forces responsible for their occurrence. It is not a distinct science, but brings together the subject matter of many earth sciences to give a general knowledge of the nature of man's natural environment. It deals with the natural aspects of geography, such as mountains, rivers and natural vegetation. The subdivisions of physical geography include the following branches.

(a) Geomorphology or Physiography

It seeks to explain the origin and development of landforms and to interpret their arrangement and distribution over the earth. This science, concerned mainly with surface features and topography, is closely related to geology. The related fields are hydrology, the study of surface and underground water, and physical oceanography, dealing mainly with ocean currents, tides, waves and under seas landforms.

(b) Climatology

It deals with the study of composite weather conditions over an extended period of time. It deals with the kinds, causes, and characteristics of climates throughout the world. Its final aim is to recognize the different types of climate and to interpret their regional pattern of distribution of the world's animals in a class called biogeography.

(c) Oceanography

The study of oceans, including the nature of the water, its movements, its temperature, its depth, marine life, marine vegetation (flora and fauna) etc. is known as oceanography. It is as important to mankind as any other branch of science. Oceans are the great sources of food, cheapest highways for transportation and main and cheap sources of energy (hydro-electric power).

(d) Mathematical Geography

It deals with the earth's size, shape, and movements. It is also concerned with the effects of heavenly bodies, such as the sun and moon, on the earth. Tides and seasons of the year are examples of such effects.

(e) Biogeography-Plant and Soil Geography

Plant geography and soil geography are concerned with the kinds and the distribution of the earth's natural vegetation and soils. Plant geography is often grouped with animal geography, the study of the distribution of the world's animals, in a class called biogeography.

II. HUMAN GEOGRAPHY

Human geography is the study of man and how he lives in his physical and cultural environment. One of the chief aims of this study is to provide a better understanding of man in the different parts of the world. Human geography deals with the man made aspects of geography such as agricultural fields, villages, railways, roads factories etc. The subdivisions of human geography include the following branches.

(a) Economic Geography

It deals with the study of how man makes living. It is especially concerned with the earth's natural resources and how they are used. Specialized phases of the study include agriculture, manufacturing and transportation geography.

(b) Population Geography

It involves the study of people, their distribution, density, vital statistics, growth, movement, occupational structure and groupings in settlements, It is an important branch of the subject.

(c) Geography of Settlements

It is a branch of geography which deals with the study of nature of human shelter, its origin and various forms. Horizontal expansion of settlements and their developments are also studied. The subject also deals with the relationship of nature of settlements and natural environment.

(d) Urban Geography

It studies the cities and towns in relation to their location, size, shape and functions.

(e) Social Geography

The social geography deals with the social and cultural aspects such as habits, food, shelter, clothing, language, religions etc. The subject deals with the study of population, urban and rural settlements and social activities.[4]

(f) Political Geography

It is the study of the influence of geography on nations, national interests and international relations. Such things as area, growth and development, boundaries and territorial size, shape and location are considered.[3]

(g) Historical Geography

It refers to geographical history. It involves a reconstruction of past environments, either as a cross-section or a series of successive sections in time or a sequential and retrospective appraisal of changes throughout time. It may include only the sources of material available at the time for which the reconstruction is being made or would allow all available material to be used. It deals with the study past geographical picture of an area which helps to understand the area of today.

GEOLOGICAL TIME SCALE

The past history of the earth is divided into major divisions, the era, which indicate the degree of evolutionary advances and distribution of the animals. The eras are divided into periods and the periods are divided into epochs. Each is characterised by advance of some and decline of other groups of animals. The eras are as under which are five in number:

1. Archeozoic Era

It is the oldest era which developed with the formation of the earth's crust, when the rocks and mountains were already existing and the processes of erosion and sedimentation had begun. It shows the huge deposits of limestone and iron ore sand traces of graphite or pure carbon. The era was probably marked by catastrophic and wide spread volcanic activities and the fossils remains are almost totally wanting, but it is presumed that life must have been abundant in Archeozoic Era.

2. Prote Rozoic Era

It is characterised by the deposition of large quantities of sediments and covers approximately one billion years of time. It has imperfect fossils of marine algae, bacteria, shelled prolozoa, jelly fishes, corals, wormtubesbranchiopds and arthropods.

3. Palaeozoic Era

It is of about 370 million years of duration. It presents for the first time the best possible fossil records. It is further subdivided into the following periods:

(a) Cambrian Period

The oldest period of palaezoic era contains fossils of almost all the present day animals pyla except the chordates. But the fossils available are of marine forms. It includes arachnid-like forms snails, trilobits, carales, sponges and echinoderms.

(b) Ordovician Period

It includes the fossils of ostracoderms, the first jawless, bottom-dwelling fishes without fins.

(c) Silurian Period

It is marked by the appearance of first land animal. The fossils of fish were abandant.

(d) Devonian Period

It marks the evolution of different types of fishes from ostracoderms. The first jawed form bony fish like Shark, lung-fishes, lobe-finned fishes, a ray-funned fishes also appeared. Hence this period is also known as "the age of fishes". By the end of this period the first vertabrates also made their appearance.

(e) Carboniferous Period

In this period amphbians and reptiles made their appearance. The different insects also appeared in this period.

(f) Permain Period

Mammals, reptiles and rapside made their appearance in this period.

4. Mesozoic Era

It lasted for about 167 million years starting from about 230 million years ago, it constitutes the "age of reptiles". Since the evolution divergence and extinction of reptiles occurred during the period, it has been divided into following three periods.

(a) Triassic Period.

(b) Jurassic Period.

(c) Cretaccous Period.

The fossils of toothed birds are available in this period. The modern birds made their appearance in this period.

5. Coenozoic Era

This era extends some 70 million years and is divided into Tertiary and Quarternary periods. It is commonly known as the age of mammals and possess man as the highest evolved mammal.

(a) Tertiary Period

During this period the mammals advanced and expanded into a varied progressive group. The primitive mammals of early tertiary period are known as archaic mammals which gave rise to modern mammals.

The fossil records of certain ungulates like horse, elephant and first apes appeared in Miocene period. Modern birds also appeared in this period.

(b) Quaternary Period

It includes the one and a million years and is sub-divided into (i) Pleistocene and (ii) Recent Period. The plestocene fossils are similar to modern mammals and birds. Sabre-toothed tiger became extinct during this period and the man appeared in recent period.

2
Solar System

The sun and the celestial bodies that travel through space in closed orbits (paths) under the sun's gravitational influence is known as solar system. In other words sun (star) with its attendant family of planets and their satellites, which revolve around it, is known as a solar system. In addition to the sun, the solar system consists of nine major planets, thousands of minor planets called asteroids, satellites, such as the moon which travel around planets comets, meteors, meteoroids, dust and gasses. The name of the system comes from "sol," the Latin word for "sun".

The surface temperature of the sun is about 10800 degree F (6000 degree c) and it increases to 36 million degree F (20 m degree c) in the centre. The fiery gasses that leap up in whirls of glowing flames are all over its surface. It is about 334000 times as big as the earth in size. The diameter of the sun is about 866300 miles i.e. 109 times greater than that of the earth. Greater the size of the steller body larger is the gravitational pull. It has 28 times more gravitational force than that of the earth. It is because of this gravitation that the various planets are kept in their orbits. It rotates on its axis once in about 25 days. Though, it is at a distance of about 93 million miles from the earth, its rays with considerable warming power reach our earth.[68]

The stars are self-luminous bodies. Because of their distance they appear very small in size. Proxima Centauri, the nearest star, is 200 thousand times more distant than that of the sun. According to an estimate there are about 160 million stars in the universe, in which about 7000 are seen by naked eye. Sir, J.E. Goar is of the idea that one can see upto 70 million stars through a telescope. But according to Prof. Newcumb Auryang, about 100 million stars can be seen with the help of a powerful telescope.

The pole star is seen in the zenith at the north pole and therefore, may be used to find out the north direction from any point on the earth's surface in the northern hemisphere. To find out the pole star, an imaginary line may be drawn through the two stars known as the pointer stars in the plough-like shape constellation or Great Bear. Then this line may be produced to about five times its length. This line will very nearly pass through the pole star. The Great Bear, a group of seven stars always revolves round the bright pole star which only rotates on its axis without changing its place in the north.

The arrangement of movements in the solar system is that the satellites revolve round the planets and the planets move around the sun. The moon is a satellite of the earth. It revolves round the earth from west to east on an elliptical path taking approximately one month i.e. 29 days and 12 hours to complete its circuit. The moon like the earth has no light of its own but shines by reflected light of the sun. The phases of the moon are the result of its position in relation to the earth and sun. When the moon is between the earth and the sun, in a straight line, the dark side faces the earth and no moon is visible. Immediately afterwards a new moon begins to appear known as new moon. When the earth is between the moon and the sun, again in a straight line, all of the moon's illuminated portion is visible, called as the full moon. When the moon is in its first and third quarters, its direction from the earth is at right angles to that of the sun. In first quarter of its orbit one half of the moon's surface is seen, known as half moon. The moon's orbit is inclined at an angle of 5 degree to the plane of earth's orbit. That is why we do not have a total solar eclipse, every time there is a new moon. The moon has a one-fourth of the diameter of the earth i.e. 2160 miles with a surface area of about 14657 thousand miles. It is at a distance of 239 thousand miles from the earth.

The sun is the chief source of heat and energy. It has its own light and energy due to thermonuclear reactions wherein hydrogen nuclei combine under intense temperature and pressure to form helium nuclei which release vast amount of energy by fission. It is the largest and most massive body in the whole solar system. Its great mass is the basis for the gravitational attraction that holds 9 major planets and all of the other bodies in the solar system in their orbits. These nine planets of the solar system can be divided into two groups.

(A) THE INNER PLANETS (TERRESTRIAL PLANETS)

1. Mercury

The smallest and closest planet to the sun in mercury, only slightly larger than the moon. One side of it is always exposed to the sun, while the other hemisphere lies in total darkness. It lies at a distance of 579.24 lakh km from the sun. Its diameter is 4,827 km. It revolves round the sun in 88 days. It has got no satellite. It is also called the "morning star".

2. Venus

The next closest planet is Venus. Observations of its surface has always been very difficult because its atmosphere is thick and cloudy. It is believed that the high temperatures experienced on the planet (470°C) are caused by a high concentrations of carbon dioxide in the atmosphere, which traps the heat of the sun. Its size is just equal to that of the earth with a diameter of 12872 km. It rotates on its axis in three days and revolves round the sun in 224 days. It has a distance of 1078.03 lakh km from the sun. It has got no satellite and is also called as "evening" star or shy planet.

3. Earth

The earth is the third planet from the sun and fifth largest planet in the solar system. It is the only planet which has large quantities of water and oxygenated atmosphere, hence it has got progressive life. It revolves round the sun in 365 days and rotates round its axis in 24 hours. Its diameter is 12872 km. Moon is the satellite of the earth which revolves round it. Earth is 93100,000 miles away from the sun.

4. Mars

The last of the four inner planets is mars. It is comprised of an endless red desert where howling winds cause massive dust storms that linger in the atmosphere for months at a time. The small amounts of water on the planet are contained in the polar ice caps, together with quantities of frozen gas. Its diameter is 7014 km. It rotates on its axis in 687 days. It is (1/9th) one ninth of the size of the earth. It has a distance of 2268.69 lakh km from the sun and has two satellites.

(B) THE OUTER PLANETS (GIANT PLANET)

5. Jupiter

Jupiter, by for the largest planet in the solar system, is composed mainly of hydrogen and helium and in many ways its structure resembles that of a star. The surface of the planet, streaked like marble, is marked by a red spot. Its diameter is 142,222 km and is eleven times greater than the earth. It rotates on its axis in 10 hours and revolves on its orbit around the sun in 12 years. It has a distance 77,721.47 lakh km from the sun. Jupiter has 14 satellites.

6. Saturn

Saturn is nearly as far away from Jupiter as Jupiter is from the sun. It appears to be composed chiefly of hydrogen and helium and has the lowest density of the nine planets. It is encircled by a cluster of rings, probably composed of a multitude of ice-covered rocks. It has a distance of 13853.49 lakh km from the sun, with a diameter of 114239 km. It revolves round the sun in 291 days. Saturn has got 10 satellites which revolve round it.

7. Uranus

Uranus and Neptune are the last of the giant planets. Its distance from the sun is 28,672 lakh km with a diameter of 55499 km. It revolves round the sun in 4 years. It has got 5 satellites.

8. Neptune

It is very remote from the sun i.e. 44907.19 lack km, hence it takes 165 years to complete one orbit around the sun.

9. Pluto

Pluto is 5925 million km away from the sun. It is visible only through the largest and powerful modern telescopes, because of its large distance from the sun. The planet is slightly smaller than mars and astronomers have recently discovered that it has a moon (satellite). Its existence came to be known in 1930. It revolves round the sun in 249 years and has a diameter of 5873 km.

We can very easily remember the arrangement of solar system by remembering a single sentence "My Very Educated Mother Just Showed Us Nine Planets". In this sentence the first alphabet of each word is the first alphabet of each planet of the solar system.

Gas, dust and solar winds are other entities found in the space of our solar system Solar winds refers to a constant stream of subatomic particles (Hydrogen and Helium) emitted by the sun which radiate throughout the solar system.

Besides the planets there are asteroids, comets, meteoroids etc. in the solar system, which will be more clear from the details below:

ASTEROIDS

Asteroids are small and irregularly shaped. Most have orbits that lie between the orbits of mars and Jupiter. There is evidence that indicates that some asteroids also have moons (satellites).

COMETS

The comets travel out from the sun farther than any other bodies in the solar system. Near the sun, comets have glowing head and long gaseous tails. Away from the sun, where little radiation strikes them, they are dark and have no tails.

METEOROIDS

Meteoroids are pieces of rock and metal that travel through space. When they intersect the earth's atmosphere they are called "meteors", which are numberless in the solar system.

Table 1

Solar System

Sl. No.	*Planets*	*Satellites*	*Period of Revolution*	*Period of Rotation*	*Diameter (in Km)*
1.	Mercury	0	88 days	59 days	4880
2.	Venus	0	224.7 days	243 days	12104
3.	Earth	1	365.26 days	23.56 Hrs.	12756

Sl. No.	Planets	Satellites	Period of Revolution	Period of Rotation	Diameter (in Km)
4.	Mars	2	686 days	24.37 Hrs.	6787
5.	Jupiter	14	11.86 years	9050 Hrs.	142800
6.	Saturn	15	29.50 years	10.40 Hrs.	120000
7.	Uranus	5	84.01 years	11 Hrs.	51800
8.	Neptune	2	164.8 years	16 Hrs.	49500
9.	Pluto	1	247.7 years	6.9 days	6000

Source: Principals of physical Geography by Dr. R.C. Baghat p. 13.

THE SHAPE AND SIZE OF THE EARTH

From years of accumulated knowledge, experience and observations in different parts of the world, it has become an established fact that the earth is a sphere. But it was first Aristotle and then Copernicus who declared the earth as round. Later on Sir, Isaac Newton by his experiments proved that it is flattened at poles and bulged at equator. It was confirmed by French mathematicians in the 18th century. They found that it has an equatorial circumference of 24897 miles and its polar circumference is less by 83 miles. In other words its equatorial diameter is 7926 miles and its polar diameter is shorter by 26 miles. The surface of the earth covers an area of about 196950 thousand sq. miles. In which 71 percent of the area is covered by water while the only 29 percent of its surface is land.

PROOFS AND EVIDENCES OF ROUND EARTH

(1) The Lunar Eclipse

The basis of the Aristotle's belief that the earth is round was the observation of lunar eclipse. The shadow cast by the earth on the moon is always circular. It is very clear that only sphere can cast such a circular shadow.

(2) The Sunrise and Sunset

As the earth rotates from west to east, the places in the east see the sun earlier than those in the west. It indicates that the earth is round. If it were flat, the whole world would have sunrise and sunset simultaneously.

(3) Spherical Planets

The telescopic observations reveal that sun, moon, satellites and stars have circular outlines. Thus it strengthens the belief that earth is a sphere.

(4) The Ship's Visibility

When a ship leaves harbour, first the hull disappears and then gradually the masts go out of view. In the same way an observer first sees the masts and then the hull, when a ship approaches the coast. If the earth were flat, the entire ship would be seen at once.

A CURVED EARTH

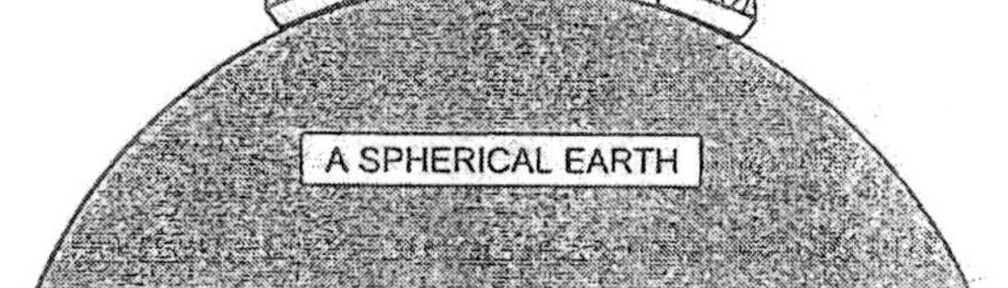

Fig. 1. The mast of a ship is seen before the hull on curved horizon.

(5) Circumnavigation

The first voyage around the world by Ferdinand Magellan and his crew, from 1519 to 1522 proved beyond doubt that the earth is sperical. Because they reached the original point from where they had started.

(6) Aerial Photographs

Photographs taken from very high altitudes through satellites and Rockets show the spherical shape of the earth. It is the latest and accurate evidence of sphericity of the earth.

EARTH'S PLANETARY RELATIONS

Our earth revolves round the sun, while the moon revolves round the earth. Hence, it is has a direct relation with both the sun and moon. The earth rotates, besides revolution, on its axis, which results in the formation of days and nights. Seasons, tides and eclipses are the other results of these planetary relations and motions.

ECLIPSES

The total or partial obscuration of the light of the sun or moon, by casting shadow of the moon or earth, is known as solar and lunar eclipse.

(a) Solar Eclipse

The obscuration of the light of the sun by moon, while passing between earth and sun is known as solar eclipse. The moon casts its shadow on the sun and it appears dark for few minutes from the earth. It occurs at a new moon only, but not at every new moon. As the moon's orbit is inclined to the plane of the elliptic at an angle of 5 degree. Hence, it takes place only when all the three are in the same plane.

The places under the umbra of the moon have total solar eclipse while the places under the penumbra of the moon experience partial solar eclipse on the earth. When the moon's shadow does not reach the earth and only a ring of light is visible round the circular disc of moon, it is said to be an annular eclipse.

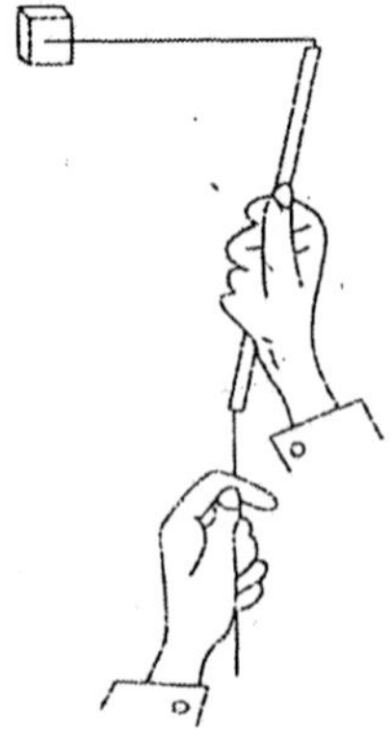

Fig. 2. To show how centrifugal force increases with decrease in distance from the axis of revolution & vice versa.

(b) Lunar Eclipse

The obscuration of the light of the moon by the earth, when it comes directly between the sun and moon, is called as lunar eclipse. It can, thus, occur only at the time of full moon, because moon's orbit is inclined at about 5 degree to the ecliptic plane. If the moon is completely obscured, there is said to be a total eclipse and if partially, a partial lunar eclipse. The lunar eclipses are far less frequent than solar eclipses. This is because at every place where the moon is above the horizon at the time of eclipse, the eclipse is visible. A total lunar eclipse may last as much as two hours.

ROTATION OF THE EARTH

The earth spins round on its axis from west to east and completes one rotation in 23 hours, 56 minutes and 4.09 seconds. It is also known as the daily motion of the earth. This motion is responsible for causing the days and nights. The earth being a non-luminous body depends on sun for its light. The sun can face only one side of the earth during rotation, which experience the day while its opposite side remains in darkness, experiences the night. The daily apparent movement of the sun and stars from east to west is also the cause of the rotation of the earth. We are able to have a sense of time e.g. morning, noon and evening also on the basis of rotation. The four directions of the earth are also fixed on its basis. It causes deflation in the direction of winds and currents e.g. towards the right in northern hemisphere and left in southern hemisphere.

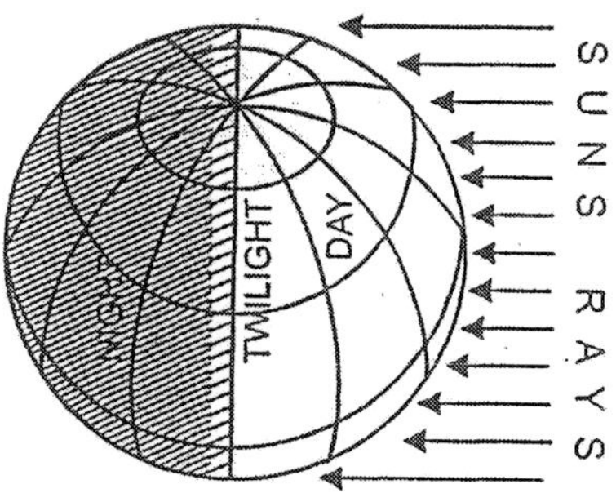

Fig. 3. Day and Night.

REVOLUTION OF THE EARTH

Besides rotating on its axis, the earth revolves round the sun on a fixed oval path known as orbit. The earth completes one revolution round the sun in 365¼ days (one year). That is why the revolution is also known as the yearly movement. During the revolution the axis of the earth remains inclined to the plane of its orbit at an angle of 66½ degree. This inclination is responsible for the variation in the length of the day and night and seasons. The main seasons i.e. winter, summer, spring and autumn are the four special positions of earth in revolution to the sun on its orbit around the sun.

As the axis of the earth always points in the same direction, the northern hemisphere is tilted towards the sun for six months and southern hemisphere during the other six months. This tilt gives continuous sun light to the north pole for six months while its counterpart remains in total darkness for the same period. For other months, reverse is the position.

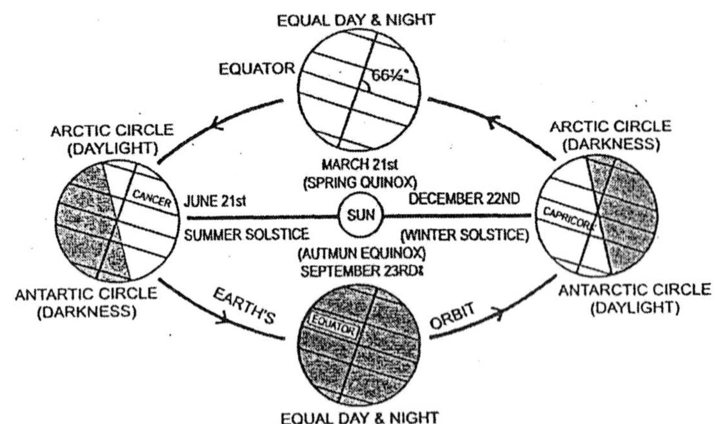

Fig. 4. The revolution of the earth and its effects on seasons and the variations of lengths of day and night.

LATITUDE AND LONGITUDE

In order to understand the position, distance and direction of a place, east and west or north and south, a set of imaginary lines has been adopted. These lines on the surface of the earth are known as latitudes and longitudes. The latitudes are the lines which run parallel to the equator, separate the sphere into unequal parts. These are also known as parallels, make small circles upto 90 degree north latitude except the equator, zero degree latitude. The most important lines of latitudes are the equator (0 degree), the tropic of cancer (23½^0 N), the tropic of capricorn (23½^0 S). the arctic circle (66½ N and the antarctic circle 66½ S. Each degree is sub-divided into 60 minutes and each minute into 60 seconds, while the distance of each degree being about 69 miles.

On the contrary, each two opposite longitudes make Great Circles, any of which divides the earth two equal parts. The longitudes also called as meridians run from pole to pole. The longitude passing through the Royal Astronomical Observatory at Greenwich, in London has been accepted as the standard or prime meridian (0 degree) in 1884 by an international agreement. From this all other meridians radiate eastward and westward upto 180 degree.

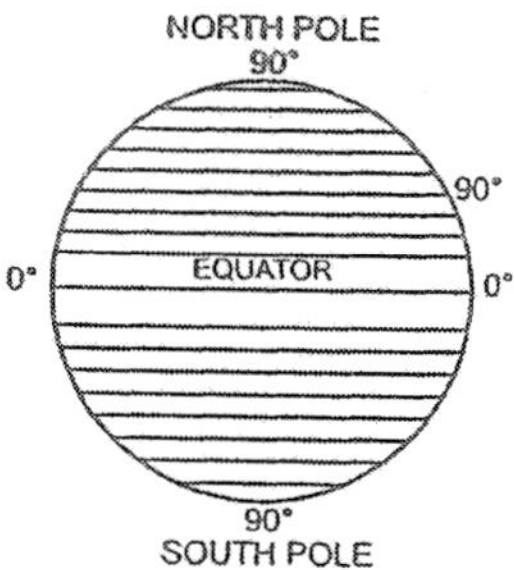

Fig. 5. Latitudes (Parallels of Latitude)

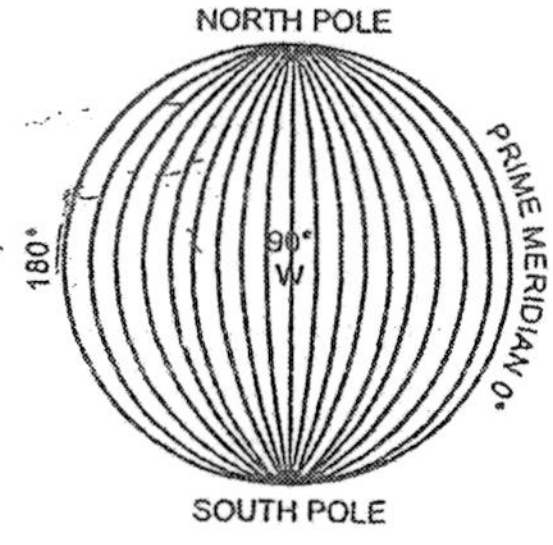

Fig. 6. Longitudes (Meridians of Longitudes)

LONGITUDES AND TIME

Since the earth makes one complete revolution of 360 degree in 24 hours, it crosses 15 degree in one hour i.e. 1 degree in 4 minutes. As the earth rotates from west to east, the local time is advanced by one hour after covering every 15 degree eastwards. Conversely, if we go westwards the local time is retarted by one hour. It means that the places east of Greenwich see the sun earlier and gain the time, whereas the places west of it see the sun later and lose the time. Hence, when it is noon, in England 0 degree longitude, the local time for Chennai (Madras) 80 degree east longitude will be 3 hours 20 minutes ahead or 5.20 pm. But for New York 74 degree west will be 4 hours 56 minutes behind England or 7.4 am. It means that when the people of England are having lunch, the Indians will have dinner and the people of New York will have breakfast.

STANDARD TIME OR ZONE TIME

The places at the same meridian have the same local time but the places at various meridians have different time. This involves a great difficulty in every sphere of life and international means of communication. To avoid all these difficulties, each country has accepted a particular meridian or meridians, the local time of which is taken as the standard time for that country or region. The India has taken 82½° E as standard meridian for the whole country which passes through Allahabad. The whole world has been divided into 24 standard time zones. The countries like USA, Canada and erstwhile USSR with large east west stretch have adopted several time zones.

INTERNATIONAL DATE LINE

The line passing along 180 degree meridian, where the date changes by one day as it is crossed, is known as International Date Line. This line has been accepted as I.D.L. to avoid the confusion of day and date by crossing it. A traveller loses a day by crossing this line from east to west, while he gains a day from west to east journey. It is only because of the loss or gain in time at the rate of one hour per 15 degree of longitude. Thus for a traveller it is mid night Friday on Asiatic side, it will be mid night Thursday for him on the American side.

The international date line in the mid-Pacific curves from normal 180 degree meridian at the Bering Strait, Fiji, Tonga and other islands to prevent confusion of day and date. The 180 degree line was chosen because it passes through the mid-Pacific, where the change of date causes the least inconvenience.

THE ORIGIN OF THE EARTH

The earth on which all types of life exists, is a member of the solar system. In comparison to other planets i.e. Jupiter, Saturn, Uranus and Neptune, the earth is very small in size. But at the same time it is the only planet, in the solar system that is suited to all types of life, as it has an atmosphere. Most of the scientists agree that the earth had its origin about ten billion years ago, in a vast interstellar cloud of dust and gas. According to the widely held belief, it collapsed in upon itself (with an unknown cause), the forces involved caused the cloud to start rotating. This rotation flattened the cloud into a disc with a dense centre. At the centre of this disc, frictional forces caused the hub to burst into a thermonuclear fire, which gave rise to the sun. Localised clumps of gas (spiral arms) were thrown off, which condensed to form the planets. This suggests why all the planets orbit the sun in roughly the same plane. It also suggests that many other larger planets like that of Jupiter and Saturn, have also formed their own solar system in the same way by ejecting the material outside.[10]

In the beginning, the nebula was moving in the form of flat disc. Due to high temperature in the centre. The heavy particles such as iron and aluminum settled down in the centre of the flat disc. Out of this material the inner planets, like mercury, venus, earth and mars were formed. Thus the inner planets are heavier than the outer planets formed of the higher gaseous particles.

During the 19th century, various hypothesis were advanced about the origin of the earth. These hypothesis can be grouped into two general classes:

(i) Monistic Hypothesis

The pioneers of this hypothesis believed that only one mass or body was responsible for the creation of the solar system. The representative of this theory is Nebular Hypothesis. The advocates and pioneers of this hypothesis were Kant and Laplace.

(ii) Dualistic Hypothesis

The advocates of this hypothesis believed that two separate masses or bodies were responsible for the creation of the solar system. The representative of this hypothesis is the Planetesimal Hypothesis and the Tidal Hypothesis. The pioneer of planetesimal hypothesis was the great English scientist, named T.C. Chamberlain and the advocate of tidal hypothesis was another English geographer named James Jeans.

(iii) Gaseous Hypothesis of Kant

This theory is based on Newton's Law of Gravitation. This hypothesis was pioneered by the famous German professor. Emmanuel Kant, Koenigsberg University Germany in 1755. He propounded that hard atoms in nature struck against each other and by their corrosion, an intense heat was caused. As these atoms began to mix with each other on account of their attraction, they began to rotate. In this way then cold and motionless cloud (Nebula) consisted of hot gaseous mass, which began to shrink in volume. This shrinking increased the speed of rotation under the influence of the centrifugal force.

A stage came when the speed of the equatorial belt of this nebula was so great, that an equatorial ring was left behind, while the main body continued to shrink. This process continued until nine such rings were formed. Each of these rings latter formed a rotating sphere revolved round the central mass. The remaining central mass in the present sun and the rings as planets.

The original process repeated in the bigger planets, giving rise to their satellites, like moon of the earth. In this way the earth as well as the whole solar system came into origin.

Kant postulated his gaseous hypothesis on the basis of new assumptions. He assumed that supernaturally created primordial hard matter was scattered in the universe (slowly rotating cloud of gas and matter comprised of very cold, solid and motionless particles). He further assumed that the particles began to collide against each other under their mutual gravitational attractions and generated random motion in the primordial matter. It also generated friction which generated heat, hence the temperature of the primordial matter started rising. Thus, the original cold and motionless cloud of matter became in due course a vast hot nebula and started rotating on its axis.

According to Kant, the random motion as well as the rate of collision among the particles increased with the increase in temperature. It gave extra impetus to the rate of rotatory motion of primordial matter and changed it from solid to gaseous state. With the continuous rise in temperature and rate of rotatory motion the nebula expanded in size. The result was that the centrifugal force exceeded the centripetal force. Hence concentric rings were thrown out from the nebula and were aggregated at a point to form planets in due course of time. The planets gave birth to satellite in the same process while the residual central mass remained as sun. Our earth one of these planets also gave such a satellite which is known as moon.[11]

CRITICISM OF THE THEORY

Kant holds greater velocity in the sun. But the fact is that due to rotation and revolution the planets have a greater velocity than the sun.

"The heavier the mass, the more the angular velocity," has been maintained by Kant, which is wrong to apply here as it all depends on size, spin and weight.

It was one of the basis assumptions of Kant's hypothesis that there was primordial matter in the universe, but he never explained its source of origin, nor could he explain the source of energy to cause random motion in it, which was cold and motionless in its initial stage.

According to Newton's first law of motion "a body remains at rest, or if in motion it remains in uniform motion with constant speed, unless or until an external force is applied on it". The particles of the matter, as assumed by Kant, were at rest and no external force was applied on them, then what was the cause for the random motion among them?

It is an erroneous statement of mechanism that collision among the particles of primordial matter can generate rotatory motion in it.

The rotatory speed of the nebula increased with the increase of its size, assumed by Kant is against the law of conservation of angular momentum, as it never changes unless an external force is applied. Thus the very foundation on which the hypothesis was based, is proved unsound and wrong.

(iv) Nebular Theory of Laplace

Pierre Simon Marquis de Laplace (1749-1827), a French astronomer and mathematician explained extensively and gave final touches to Kant's hypothesis in 1785. In his work 'Celestial Mechanics', Laplace showed that the movements of the solar system were in conformity with Newton's theory of gravition.[12]

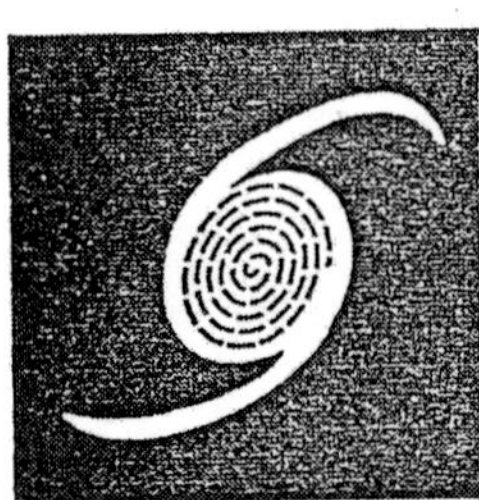
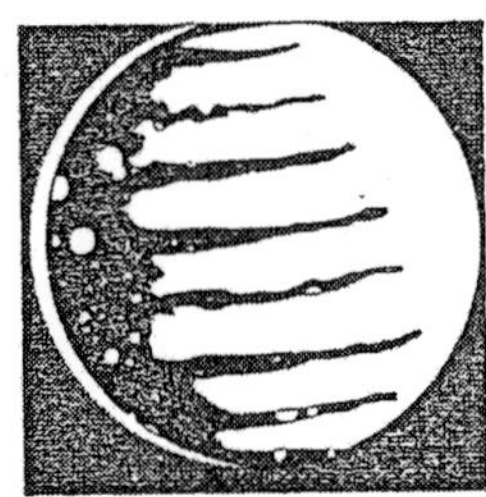

Fig. 7. Nebula and the cooling of the Earth.

He advanced that an extensive gaseous mass occupied most of the universe. This mass of gas was rotating, to which he called the sun. It was rotating at a terrific speed, due to which its outer portion began to cool down by radiation process, followed by contraction and shrinkage. This contraction accounted for a reduction in the angular velocity, hence rotation became more rapid. The outer portion being more dense was unable to keep pace with the central core and hence rings were thrown out. These rings continued to rotate in their places and revolve round the main nebula. These rings condensed into planets after some period and the remaining central mass continued as sun. The larger planets gave birth to their satellites with the same process and earth also threw one such ring, which condensed into moon.

Laplace's hypothesis is just the modified version of Kant's hypothesis. In fact Laplace postulated his hypothesis after removing the inherent weak points and errorneous concepts of Kant's hypothesis which suffered from three basic defects e.g.

(a) Large amount of heat cannot be generated due to the collision of cold particles of primodial matter.

(b) Mutual collision of particles cannot generate motion in the primordial matter and the random motion of the particles cannot generate circular motion (rotatory motion) in the primordial matter.

(c) The angular velocity of rotatory speed of the nebula cannot increase due to the increase in the size of the nebula.

In order to remove these defects Laplace assumed that:

(a) There was a huge and hot gaseous nebula in the space.

(b) This huge and hot nebula was rotating in the universe on its axis from the very beginning.

(c) The nebula was continuously cooling due to loss of heat from its outer layer by radiation process and it was also reducing in size due to construction on cooling. Thus reduction in the size and volume of the nebula increased the circular velocity (rotation) of the nebula. Hence the inner mass could not keep pace with external layer or centrifugal force exceeded the centripetal force. The result was that the rings were thrown out from the main nebula, which started rotating and moving around the nebula. These rings condensed into planets. They also threw such rings which condensed into satellites, such as earth gave a moon.

Criticism of the Theory

The nebular theory, widely accepted for hundreds of years, has several serious flaws. In the early 20th century, it was rejected and planetesimal hypothesis became popular. William Hobbes, an American geologist, says that this hypothesis is responsible for giving us not only a false conception of the nature of our planet but also of the origin of lava.

The other defect in the theory is that rings cannot condense into planets. Gaseous ring can condense into a large number of steller bodies all along the rings as the ring of Saturn.

The most serious concerns the speed of rotation of the sun. When this theory is worked out mathematically on the basis of the known orbital momentum of the planets, it predicts that the sun must rotate about 50 times more rapidly than it actually does.

Sir James Jeans propounded that the homogeneous mass will throw spirals, as a result of rotation inspite of rings.

(v) Tidal Hypothesis of Jeans and Jeffrey

Sir James Jeans, a British scientist, propounded his tidal hypothesis about the origin of the earth in the year 1919 while another British scientist, Harold Jeffrey suggested modifications in it in 1929.

According to this theory there was a huge and extensive gaseous mass wandering in the universe. In its journey through space another stellar body approached the sun and came very close to it. Both the stars exerted gravitational pull on each other. Since the approaching star was bigger than the sun, it exerted more gravitational pull and a tide occurred on the surface of the sun. As the bigger star came nearer, the height and size of the tides went on increasing. In course of time protuberances of material jutted out towards the approaching star. The two stars came very close to each other, hit and the approaching star ran away. Hence the arms of material jutted out of the sun were left behind. These spiral arms condensed into planets in due course of time and began to rotate round the sun. This theory is also known as "hit and run" theory.

Later on due to the gravitational pull of the sun there was tidal action in these planets too and the satellites were formed.

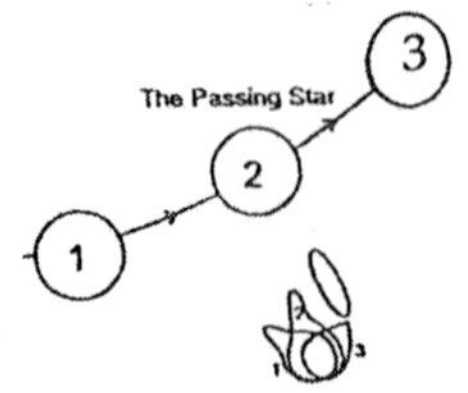

Fig. 8. The Tidal Action on Sun, according to Jeans & Jeffrays. The passing star was several times larger than the sun. In position 1, filament is small & the sun larger than in the successive positions, 2 & 3 when the filament has increased in dimension.

POINTS IN FAVOUR OF THE HYPOTHESIS

1. The tidal protuberances jutting out from the sun should have been cigar shaped i.e. thick in the middle and thinner at both the ends. The arrangement of the planets in the solar system is also like a cigar. From the first, mercury, the size of the planets goes on increasing till Jupiter the biggest in the middle and after it the size goes on decreasing till Pluto, the last and smallest.
2. The arrangement of the satellites confirms the truth of this theory, as observed in Jupiter and Saturn which have the highest number i.e. 14 and 15 satellites respectively, is also cigar shaped.
3. The theory implies that the bigger planets remained in the gaseous state for a longer time, hence they have larger number of satellites. On the other hand the smaller planets cooled early hence they have either less or no satellites. This theory is also known as Catastrophic theory or Tidal Action theory or Hit and Run theory.

MODIFICATION BY JEFFREYS

Harold Jeffrey, a British scientist, modified the original tidal hypothesis of James Jeans in 1929 and presented his concept as "Collision Hypothesis".

According to Jeffrey there were three stars in the universe before the origin of our solar system. First was the "primitive sun", the second as its "companion star", and the third was the "intruding star". The intruding star was moving towards the companion star, ultimately it collided with it and the companion star was completely smashed and shattered. Some shattered portions were scatered in the sky while the remaining debris started revolving around the primitive sun. However, the impact of collision and explosion enabled the intruding star to clear itself off from the gravitational attraction of the primitive sun and gradually vanished in the universe. The planets and the satellites were formed from the remaining debris of companion star.

Jeffrey suggested modifications in the tidal hypothesis of Jeans with intention to remove major inherent weak points of the tidal hypothesis so that it can withstand the criticisms of the modern scientific world.

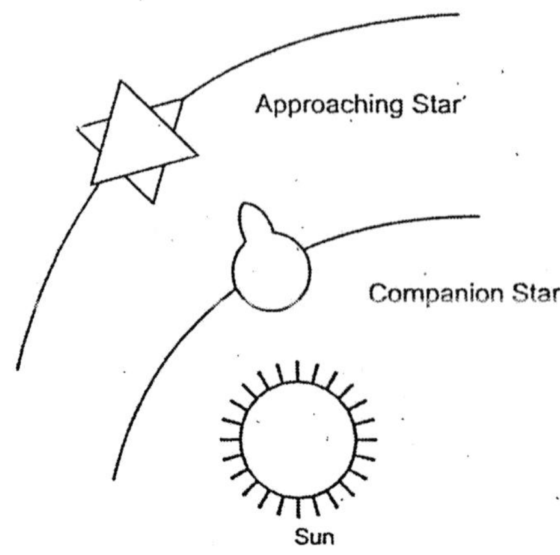

Fig. 9. Formation of plannts according to Tidal Hypothesis in the Fig.

CRITICISM OF JEANS AND JEFFREY'S THEORY

1. Collision is not possible according to B. Levin as he says that the universe is infinite in space and time and the stars are so distant from each other that the close encounter is never possible.
2. The theory does not explain that where from the intruding star came and where it went away, which caused tidal action on the sun.
3. In fact, according to this hypothesis the planets should have been very close to the sun but in reality they are far away from the sun.
4. According to the theory the satellites were formed from the tidal effect exerted by the primitive sun on the newly born planets. Here a question arises, why no satellite was formed from Mercury and Venus though these planets were nearest to the sun. In fact the sun should have exerted maximum pull on these two planets.

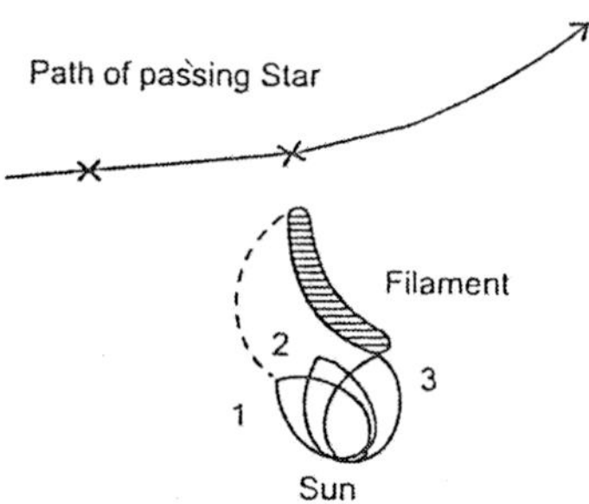

Fig. 10. Tidal disruption of Sun. Jet or ailament of gaseous matter detached from sun. (After Jeans).

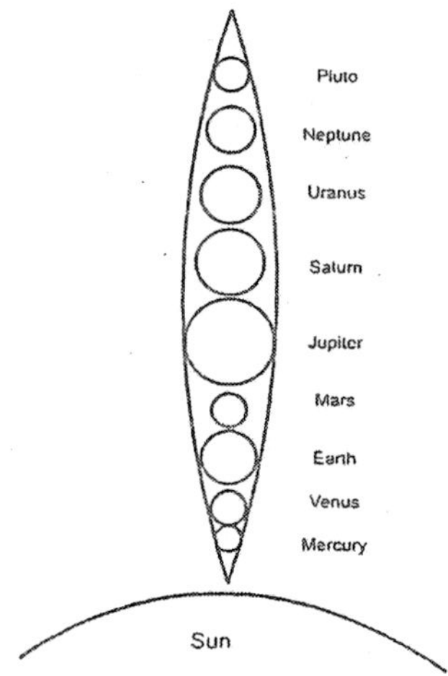

Fig. 11. Showing Cigar-Shaped arrangement of the planets of the solar system.

(vi) Planetesimal Hypothesis of Chamberlain and Moulton

Planetesimal, a theory of the origin of the solar system proposed by Forest Ray Moulton and Thomas C. Chamberlain in about 1900—1905 (both belong to the university of Chicago U.S.A.). This theory states that planets were formed by the accumulation of extremely small bits of matter-planetesimal, that revolved around the sun. This matter was produced when a passing star almost collided with the sun. During the near collision, hot gases were pulled out of both stars and the gases then condensed. The planetesimal hypothesis was widely accepted for about 35 years.

The greatest flaw in the theory is the assumption that the material drawn out of the stars would condense. The extremely hot gases that make up a star are held together by the gravitational forces within the star. Once the material was pulled away to where the gravitational forces were weaker, it would expand because of its heat. Before condensation could take place, the gases would have almost entirely dissipated. The planetesimal hypothesis is no longer considered a likely explanation of the origin of the solar system.[13]

According to Chamberlain, initially there were two heavenly bodies (stars), in the universe i.e. Proto sun and its companion star. The Proto-sun was formed of very small particles which were cold and solid. The introducing star came very close to the proto-sun and infinite number of small particles were detached from the outer surface of the proto-sun due to the massive gravitational pull of this giant star. Chamberlain termed these particles as planetesimals.

Gradually, large planetesimals started attracting small ones and ultimately they grew in the form of planets due to continuous accretion of infinite number of planetesimals. The remaining proto-sun changed into present day sun. The satellites of the planets were created due to the repetition of the same processes and mechanisms.

According to this theory the main force of ejection of small planetesimals was the tidal force exerted by the intruding (approaching) star on the outer surface of the proto-sun.

Chamberlain and Moulton's theory is slightly different from that of Jeans and Jeffreys, as according to this theory the sun was periodically emitting eruptions due to some unknown cause. The intruding star came near it and produced tidal action over and above these eruptions, so they became long. When it receded, the eruption arms became bent and spiral (spiral nebula). The bigger eruptions gave planets while the smaller ones gave satellites. These planets started revolving round the sun alongwith the satellites.

CRITICISM OF PLANETESIMAL HYPOTHESIS

1. According to the planetesimal hypothesis of

Chamberlain, the size of the planets was dependent upon:

(a) The amount of accretion and aggregation of planetesimals around the nucleus.

(b) Amount of planetesimals available in the particular orbit.

(c) The attractional force of so-called nucleus of the planets.

Thus, the nucleus of any orbit would have acquired and accreted any number of planetesimals and would have gained any size. Thus the planets should have not been arranged in any order such as the present cigar-shaped solar system. The increase in the size of the planets away from the sun continues upto Jupiter and then the size goes on decreasing upto the last planet Pluto, the smallest and farthest. Chamberlain, thus, does not offer any explanation of this type of arrangement.

2. According to this hypothesis, the planets always remained in solid state, whereas the others say that they were initially in gaseous and liquid state and gradually condensed into solid state.
3. Such a close encounter between the stars is not possible, in the infinite space of universe, as shown by Chamberlain.
4. It presupposes eruptions before the approach of star.

(vii) Binary Star Hypothesis of Russell

H.N. Russell, an American astronomer, propounded his hypothesis in 1937 to remove the shortcomings of tidal hypothesis of James Jeans. Russell says that there were two stars near the primitive sun in the universe. In the beginning the companion star was revolving around the sun. Later on an approaching star came near the companion star but the direction of revolution of the approaching star was opposite to that of companion star. It was believed that the distance between two stars might have been about 4800 thousand to 6400 thousand km. It means that the approaching star might have been at a far greater distance from the primitive sun. Thus, there was no tidal action on sun but on the contrary large amount of matter of the companion star was attracted towards the giant star because of its massive gravitational pull. The ejected matter started revolving in the direction of giant approaching star. Later on planets were formed from this ejected material. In the beginning the planets were nearer to each other and hence due to mutual attraction some matter ejected from planets, which formed the satellites.

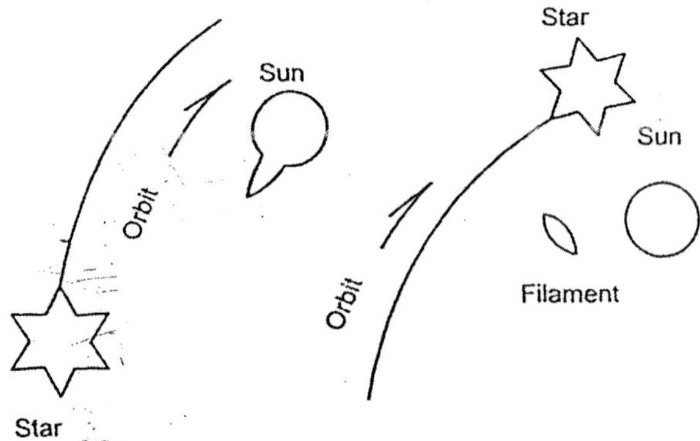

Fig. 12. Showing Origin of earth according to Binary Star Hypothesis of H. Russel.

CRITICISM OF THE THEORY

1. Russell did not throw any light on the fate of the remaining portion of the companion star. What happened about the residual part of the companion star, he could not answer this question.
2. According to Russell, the far off planets were brought under the gravitational domain of primitive sun after the giant star receded or disappeared. But on the other hand residual companion star, which was nearest to the sun could not come within the gravitational field of sun, Russell was unable to resolve this contradiction.
3. He did not elaborate the process and mechanism through which the planets, after their formation, were brought within the gravitational field of the sun.

(viii) Binary Star Theory of Lyttleton

According to this theory the sun was already in the universe along with a companion star. A third star (approaching star) came very close to companion star. As a result tidal protuberances jutted forth and they were so near that both the protuberances were joined. A ribbon of filament was thus formed. When the third star ran away the companion star also disappeared. The middle portion of the ribbon was captured by the sun, which began to revolve round the sun and plants condensed out of it.

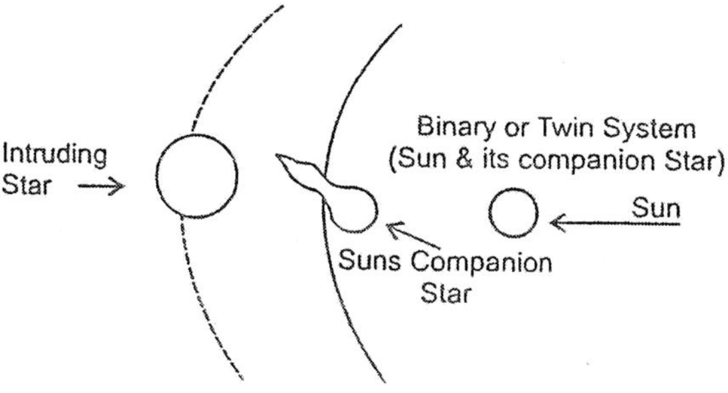

Fig. 13. Showing Lyttleton's hypothesis of Binary Star (After W.M. Smart). The planets originated from matter pulled out by the tidal action of the passing star on sun's companion.

CRITICISM OF THE THEORY

This theory has been criticized by Layton and Dr. Bhatnagar.

1. Lyton maintains that the sun would also be incised by the third star. So it was a triple star theory, but it is not so according to Lyttleton.
2. Dr. S.S. Bhatnagar says that the stars would collide as they approached very near. There would be direct collision and hence the question of the formation of planets does not arise at all. Thus it is no improvement on the theory of Jeans, rather it makes it more complex.

(ix) Rotational and Tidal Theory of Rossgunn

It tries to combine the basic principles of the theories of Laplace and James Jeans. It thinks of a rapidly rotating star almost reaching the breaking point. At this stage another star came. Both were deformed by tidal action. Then they separated, leaving a mass of tidal protuberance behind. This mass condensed into planets.

Through this theory, Rossgunn tried to explain the velocity by giving a quantitative rotation to the original star i.e. sun. But the coming of another star at such a stage is impossible. According to A.C. Bannerji, this is a far fetched way of explaining the velocity.

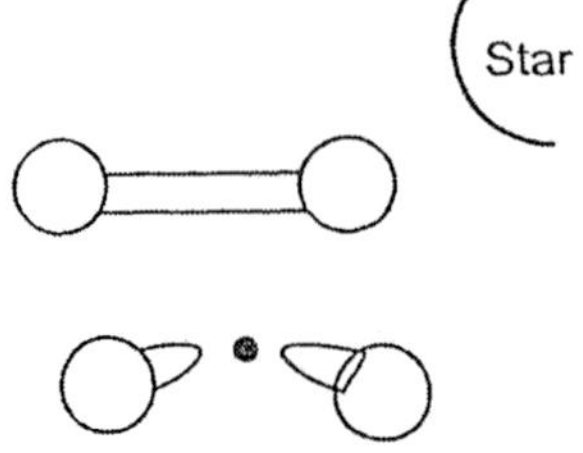

Fig. 14. Rossgunn's Tidal Theory.

(x) Cepheid Theory of Prof. A.C. Bannerjee

There is a star in the constellation cepheid and its brightness changes regularly so that maximum and minimum alternate with each other. This regular change is due to the contraction and expansion which is almost regular.

In the oscillating stage another star approached at a moderate distance and the material is thrown out at a distance. When it began to condense, the sun and the cepheid separated. The sun went one way and the cephied another. The sun stole away two fifth of the energy from the planet cepheid and this is confirmed by mathematical findings which show the sun to have twenty times more energy than the planets.

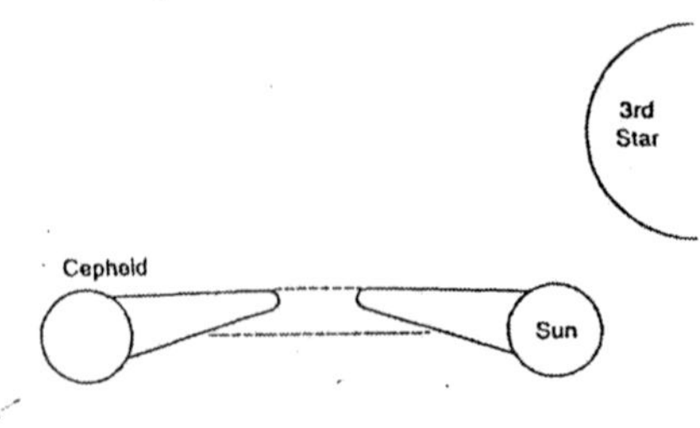

Fig. 15. Showing Cepheid Theory of Prof. A.C. Bannerjee.

(xi) Inter-Stellar Dust Hypothesis or Gas-Dust-Cloud Theory of Otto Schmidst

In 1943, a Russian geographer named Otto Schmidst pretended gas-dust-cloud theory about the origin of the earth. This theory was most favoured and universally accepted. He believes that the planets and satellites made of the swarm of diffused matter which was composed of gas and dust particles. These particles revolved round the sun under its gravitational force. They gradually contracted and assumed the shape of the flat disc. Inelastic particles of this cloud consolidated and created heat among them. The gas-dust-cloud is also known as nebula.

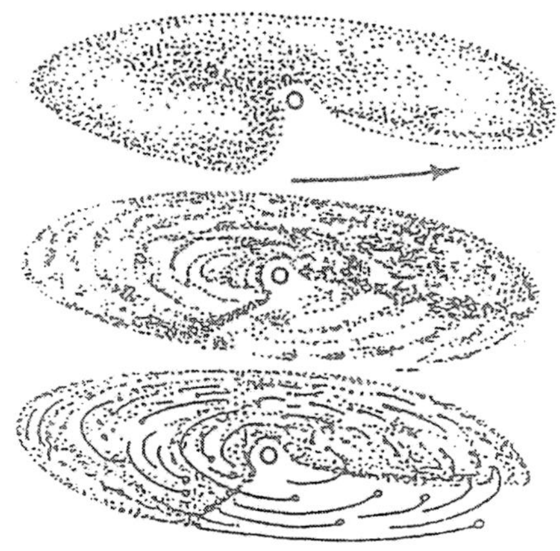

Fig. 16. Formation of disc of dark matter and embryos of the planets according to Schmidt.

The speed of the nebula increased with its reduction of size. The angular momentum of the nebula was redistributed forming planets and satellites.

The special characteristic of this theory, which goes in support of it is that the inner planets are composed of heavier materials as they consolidated in the inner part of the disc. Hence they have low angular momentum and smaller, orbits. The outer planets are composed of light gaseous materials, hence they have greater angular momentum and larger orbits.

The scientific researches about the universe have given ample evidence of the presence of "dark matter" in the form of gas and dust particles known as "gas, dust and cloud", in the universes. Though Schmidst did explain the mode of origin of these dark matter but it may be safely assumed that these gaseous clouds and dust particles might have been formed from the matter coming out of the stars and meteors.

According to this theory our sun during its "galactic revolution" captured the dark matter of the universe, which started revolving round the primitive sun. Collision among the dust particles started the process of aggregation and accretion around the bigger particles which became the embryos of the future planets. But gas particles could not condense as they could not be organised due to their continued motion. The embryos captured more and more matter-to grow in size to become asteroids. These asteroids further grew in size to become planets.

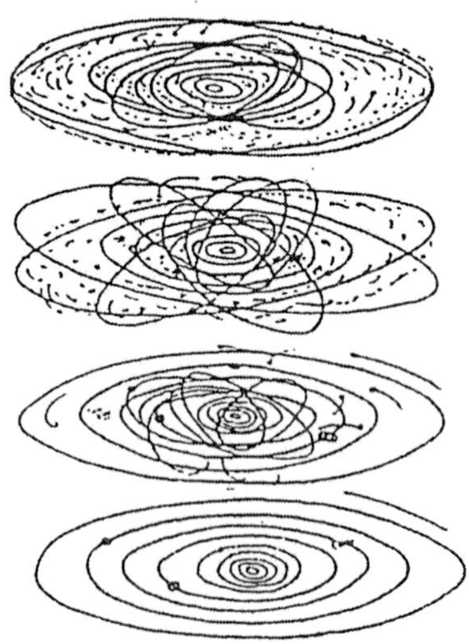

Fig. 17. Final stage of the formation of planets according to Schimidt (after B. Levin).

(xii) Supernova Hypothesis of Hoyle

F. Hoyle, a mathematician of Cambridge University, U.K.

presented his theory known as supernova hypothesis in the year 1946. According to Hoyle, initially there were two stars, the primitive sun and the companion star. The latter was of giant size which became supernova due to nuclear reaction. With the passage of time all the hydrogen nuclei of the companion star were consumed in the process of nuclear reaction and it collapsed and violently exploded.

The violent explosion of the companion star (supernova) resulted into the spread of enormous mass of dust (circular moving disc) which started revolving around the primitive sun. Hoyle maintained that while explosion of the companion star the nucleus was thrown out of the gravitational pull of primitive sun. This matter of the disc became the future planets.

It may be pointed out that the explosion of supernova generated intense heat enough to start the process of nuclear fusion, which became responsible for the formation of heavy elements e.g. helium, carbon, oxygen, silicon, nitrogen etc.

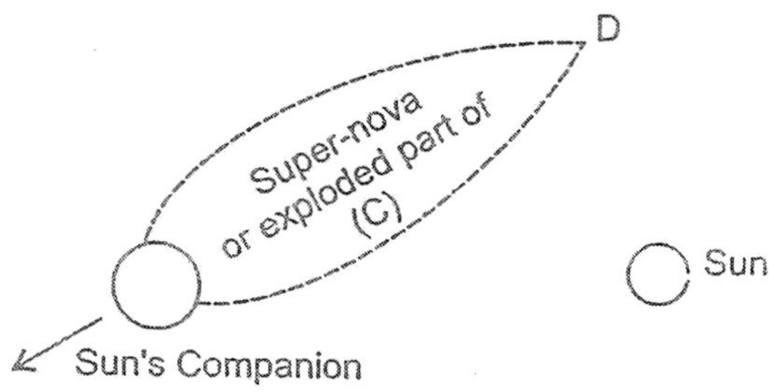

Fig. 18. Showing Hoyle's theory of supernova (After W. Smart) The direction of recoil in which the sun's companion C disappears after explosion in the direction CD.

CRITICISM OF SUPERNOVA

If the supernova hypothesis on one hand solves the intricate problems of the solar system. It fails to explain the peculiar arrangement of the planets on the basis of their size, similar direction of rotation as well as the plane of revolution and path of the planets and the lighter constituent elements of the planets of the outer circle of our solar system on the other hand.

(xiii) Big Bang Theory

According to this theory, all the matter in the universe was originally concentrated in one vast lump, hundred million light years across, called the primeval atom. Ten to twenty thousand million years ago this atom exploded and the universe started its balloon like expansion, because the matter was flung in all directions throught space. Stars formed from the matter travelling through space after the explosion of the primeval atom. The universe is still expanding. The galaxies are like spots on a balloon which is being inflated. The distance between the galaxies is increasing all the time.[17]

The Big Bang theory postulated in 1950's and 1960,s and validated in May 1972 through convincing evidences received from Cosmic Background Explorer (COBE) explains the origin of universe. The universe contained many millions of stars and each star having numerous planets around them.

There were already wispy clouds of matter stretching across vast distances. As those clouds collapsed in upon themselves, pulled together by their own gravity, they would have broken up and formed clusters of galaxies breaking up into stars like those of Milky Way. The stars might have broken up, to form their planets as our earth.

(xiv) Steady State Theory

The main rival to the Big Bang Theory is the Steady State Theory. This theory suggests that the universe as a whole had no definite beginning and will have no definite end. Individual stars and galaxies develop and die out, but new ones are continuously being born to replace them. The amount of space between existing galaxies is increasing but new matter is being created in the space to balance the total amount. According to this theory the universe has no definite size. It stretches out to infinity.[18]

(xv) Electromagnetic Theory of Hannas Alfven

In 1942 Hannas Alfven assumed that in the past the sun was rotating faster than now. It then happened to plunge itself into a nebula in which the atoms were originally electrically neutral. As the atoms were pulled by the gravitational attraction of the sun, they were ionized or electrically charged. Such ionization extended in the sun's envelope of atoms upto planetary distances.

Then Alfven applied the behaviour of electrically charged particles in a magnetic field and showed that matter would gather larger in the equatorial plane of the sun at such distances which can now be compared with the distances of Jupiter and Saturn. This matter would revolve round the sun and one of its effects would be to retard the rotation of the sun.

Later the gaseous and other atoms condensed into large planets. It is believed that Jupiter having its own magnetic field behaved in the same way as the sun with regard to the surrounding matter and Jupiter's satellites were formed.

The magnetic field of the sun is the basis of this theory. The electromagnetic force is for more powerful than its gravitational force. The ratio of former to the latter on a portion in the orbit of the earth is 60,000 to one. It is 250 to one on a proton in Pluto's orbit.[16]

But there are lacunae in the theory. One of these is that it does not explain the origin of the inner planets (mars, earth, venus and mercury).

(xvi) Dust Gas Cloud Theory of Weizsacker

The German physicist Carl Van Weizsacker showed in 1943, that all the old objections against the Kant-Laplace hypothesis can be easily removed.

This he could do on the basis of new knowledge about the chemical composition of the cosmic matter, which shows that the chemical composition of the earth was different from that of the stellar bodies. The proportion of light gases like helium and hydrogen is very small in the earth and that the sun was composed mostly of hydrogen and helium and only one percent of earth's constituents.

According to Weizsacker the gases (hydrogen and helium), dust particles as iron oxides, compounds of silicon, water droplets and ice crystals was outside the sun as rotating envelope. The collision of dust particles and their accumulation into larger aggregates resulted into planets.

According to Weizsacker, the time of formation of planets from the fine cosmic dust was about 100 million years. The same rule holds good for the satellites of the planets suggesting that the satellites originated more or less in the same way as the planets.[14]

Of the envelope surrounding the sun, the gases formed about 99 percent. The dust particles were aggregated into larger lumps to form the planets and their satellites and the asteroids, but the gases mostly hydrogen and helium escaped into interstellar space in a period of some 100 million years.

Weizsacker's theory envisages that what happened with sun must have occurred in the case of other stars also and numerous system of planets must have formed.

Thus, there must be millions of planets in our galactic system along with physical conditions resembling those of earth. No wonder, therefore, that life in its highest forms existed in the planets of our galaxy.[15]

The most inhabitable planets of the solar system are mars and venus. We may know in future about their habitability by means of space ships.

RECENT THEORIES

Among the recent hypothesis and theories of the origin of the solar system the significant are the following:

1. Fon Vaischer's Hypothesis;
2. Alfven's Inter - Steller Cloud Hypothesis;
3. Rossgunn's Rotational and Tidal Hypothesis;
4. Cepheid Hypothesis of AC. Bannerjee (of Allahabad University);
5. Kuiper's Hypothesis 1949;
6. Voittkeyich's Proto - Planetary Choudrule's Concept 1971;
7. Fosenkov's Globule Concept 1951;
8. Jupiter - Sun Binary System Hypothesis of E.M. Drobyshevski 1974.

The latest ideas of the modern astrophysics indicate that there was initially a big rarified cloud of interstellar matter which started rotation around its axis. Later on, at the stage of redistribution of matter in the rotating disc of cloud of matter maximum mass was accumulated in its central part which increased the luminosity of the core (central part) of the disc. This core of the disc became our primitive sun. Relatively lesser masses of the outer zones of the disc became planets after being condensed. This is the summary of the latest "Nuclear Disc Model," of the origin of the sun and its planets. The formation of the planets of different constituent materials and size is explained with the help of the "Chemical-Condensation-Sequence Model" of the American scientists.

3

The Interior of The Earth

If you were to go down a very deep hole, such as the shaft of a mine, taking with you a thermometer, you would find that the temperature would rise at the rate of 1°C for every 33m (1°F for every 60 feet) you went down. By the same rate, if you had gone 3 km (2 miles) down, you would have reached the temperature at which water boils (100°C). It means that, if the temperature goes on rising at the same rate, the rocks of which the earth is composed, will be in molten state and gas at the centre of the earth.

It is decidedly true it is very difficult task to have accurate knowledge of the constitution of the earth's interior, because it is beyond the range of direct observation by man. Therefore, for understanding the interior of the earth we have to depend on an indirect evidence. We have to study the earthquake waves and on getting them back take a hint of what they have passed through. This study is done through a branch of geography known as "seismology" which has helped to have some authenticated knowledge about the mystery of the earth's interior. Besides seismology there are other sources also, which provide knowledge about the interior of earth. These sources can be grouped in three divisions, the detail of which is as under:

I. ARTIFICIAL SOURCES

The artificial sources may be further sub-divided into three dimensions e.g. temperature, density and pressure.

(a) Temperature

It has been observed on the basis of information of bore holes and deep mining that temperature increases with the depth at the rate of 1°C for every 33 meters. It is evident that high temperature of 1000°C at the depth of 43 km in the technically active areas is nearer to initial melting point of the rocks of lower crust and mantle mainly basalt and peridotite.

The temperature of the upper part of the magma slab representing the upper portion of the oceanic crust has been estimated to be 0°C while the temperature of the lower part of it remains 1200°C nearer to melting point. If we calculate the temperature at the depth of 2900 km at the aforesaid rate, it should be around 25000°C at which most part of the earth would have melted, but this has not so happened. It is because the most parts of the radioactive minerals are concentrated in the uppermost layer of the earth. This fact explains the situation of high temperature in the continental crust, as disintegration and decay of these minerals generate more heat in the crustal areas.

The temperature at the depth of 400 and 700 km has been estimated to by 1500 degree C and 1900 degree C respectively. The temperature at the junction of mantle and outer molten core standing at a depth of 2900 km is about 3700 degree C. The temperature at the meeting point of outer molten core and inner solid core at a depth of 5100 km is about 4300 degree C.

(b) Density

It is commonly believed that the outer thinner part of the earth is composed of sedimentary rocks, the thickness of which ranges between half a mile to one mile. Below it is a second layer of crystalline rocks, the density of which ranges between 3.0 to 3.5. The average density of the earth is about 5.5. The density increases with the depth so it is 11.0 at core of the earth. Since 1950 several attempts are being made to calculate the density of the earth on the basis of satellites. The satellite studies have revealed the following results about the density of the various parts of the earth. The average density of the earth is 5.517g cm^3. The average density of the earth's surface is 2.6 to 3.3g. cm^3 and average density of the core of the earth is 11 gram cubic cm.

(c) Pressure

We can understand very easily, the meaning of pressure if we understand the reason of increasing density with depth (highest in the core). It is believed that the high density of the core is because of heavy pressure of overlaying rocks. It is common principle that pressure increases the density of rocks. Hence, the high density of the core of the earth is due to the very high pressure prevailing there because of superincumbent load.

II. THE THEORIES OF THE ORIGIN OF THE EARTH AS A SOURCE

Almost all the pioneers of various hypothesis and theories about the origin of the earth have assumed the original form of the earth to be solid, liquid or gaseous. According to the "nebular hypothesis" of Laplace the core of the earth should be in liquid state. Zoeppritz and Ritter have opinioned that the core of the earth is made of gases but this concept may not be accepted because if we assume so, many more problems will emerge. There may be only two possibilities viz. either the core may be in solid state or liquid state.

According to the "tidal hypothesis" the core of the earth should be in liquid state because the earth has taken its birth from the tides ejected from the primitive sun. In other assumption according to the "planetesimal hypothesis", the earth was originated due to accretion and aggregation of solid dust particles known as planetesimals. Hence according to this concept the core of the earth should be in solid state..

III. NATURAL SOURCES

The natural sources may be sub-divided into volcanic eruption, earthquakes and seismology.

(a) Volcanicity

During the volcanic eruption the hot and liquid lava spreads over the surface of the earth, is the clear indication of the molten or liquid zone beneath the surface of the earth. It is also termed as "magma chamber", which supplies the lava during volcanic eruption through a fissure or hole on the earth's surface, with a great pressure.

(b) Evidences of Seismology

Seismology is the science which studies various aspect of seismic waves generated during the occurrence of earthquakes. These waves are recorded with an instrument known as seismograph. It may be kept in mind that seismology is the only source which provides us authenticated information about the composition of earth's interior.

The place of the occurrence of an earthquake is called "focus" and the place which experiences the seismic event first is called the "epicenter", which is always perpendicular to the focus point on the surface of the earth. The focus is always inside the earth, the deepest ever measured focus is at the depth of 700 km from the earth's surface. The different types of tremors and waves generated during the occurrence of an earthquake are called "seismic waves", which are generally divided into three categories e.g. primary waves, secondary waves and surface waves.

By finding how long it takes these waves to travel to places at various distances, we can get some idea of the nature of the material they have passed through. We can even say what part is solid and which is molten.

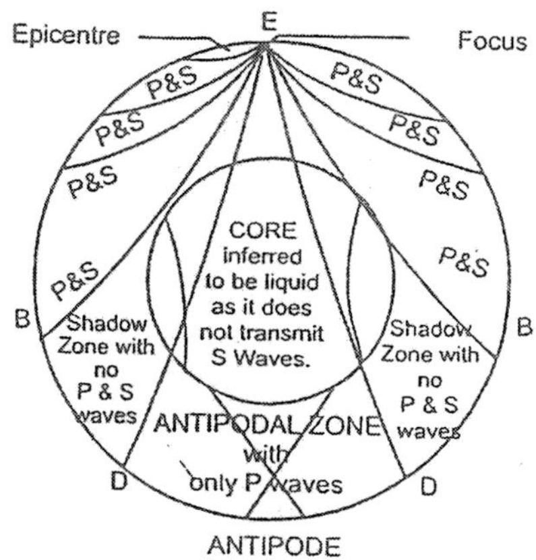

Fig. 1. Showing a section across the earth showing seismic waves (after Holmes). The diagram shows the paths of seismic waves across the earth from the focus, the shadow zone, the core & the antipodal region.

1. PRIMARY WAVES

These waves are also known as main or longitudinal or compressional or push (P) Waves. The P - Waves are fastest earth-quake waves through solid rocks. In these the particles of matter through which they pass move backward and forward in the lines of propagation of the wave. They are analogous to sound waves. Their speed is retarded through liquid material.

2. SECONDARY WAVES

These waves are also known as transverse or distortional or shake (S) Waves. The S-Waves are the waves in which the motion of particles are at right angles to the direction of prorogation. They are like waves of light or ripples of water. S-Waves cannot pass through liquid materials.

3. SURFACE WAVES

Surface waves are also called as longitudinal or long period (L) waves. The L-Waves travel through the surface regions of the crust and do not go into the interior of the earth. These waves are slowest but are marked by the largest oscillations. They are the source of destruction in major earthquakes. The speed of S-Waves is about six-tenths that of P-Waves.

From the observations of movement, speed and time taken by these waves, it has been found that higher the density of the medium through which the waves are passing, the greater will be the speed of the waves. The results of the studies or earthquake waves reveal that up to a depth of 1800 miles down from the surface, the speed of the waves goes on increasing progressively. At a depth of 1800 miles, the speed of the waves changes in a peculiar manner. The transverse waves are unable to penetrate any further. As a matter of fact the secondary waves cannot enter liquids, so it is assumed that the core of the earth must be in liquid state. The primary waves are able to enter the core but their speed is retorted to the minimum.

The geographers and geologists have concluded from the observations, of aforesaid waves that at the depth of 1800 miles we reach liquid.

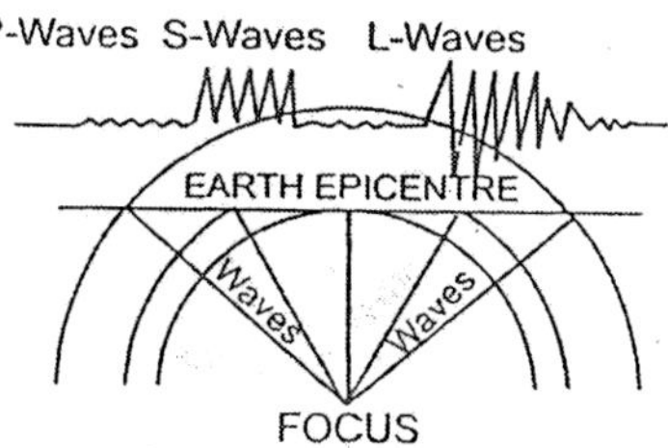

Fig. 2. Earthquake Waves.

STRUCTURE, COMPOSITION AND LAYERS OF THE EARTH

The surface of the earth is composed of about 90 chemical elements in various combination. The most abundant of these elements are oxygen and silicon.

Scientists are not completely certain about the structure and composition of the earth's interior because they cannot make direct observations. The deepest holes man has been able to drill into the earth go down only about five miles. However, by studying shock waves produced by earthquakes and nuclear blasts, scientists can estimate the density of various parts of the earth's interior.

The nature and properties of the composition of the interior of the earth may be successfully obtained on the basis of the study of various aspects of seismic waves. The recorded seismic waves denote the fact that these waves seldom follow straight paths rather they adopt curved and refracted paths. Thus, it is obvious that the earth is not composed of hogogeneous materials rather there are variations of density inside the earth.

Hence, on the basis of seismology it can be inferred that there are three distinct zones or layers of varying densities inside the earth, if we proceed downwards from the crust of the earth. These three layer or zones of the earth have been called by different geologists by different names. The German scientist Grachit has called them SIAL, SIMA, NIFE, while another scientist Jefferey calls them the Top Layer, the Middle Layer and the Lower Layer, Prof. Halmes call them the curt, the substratum and the core, Daly calls them as Outer Zone, Intermediate Zone and Central Zone. These are also known as crust, mantle and core.

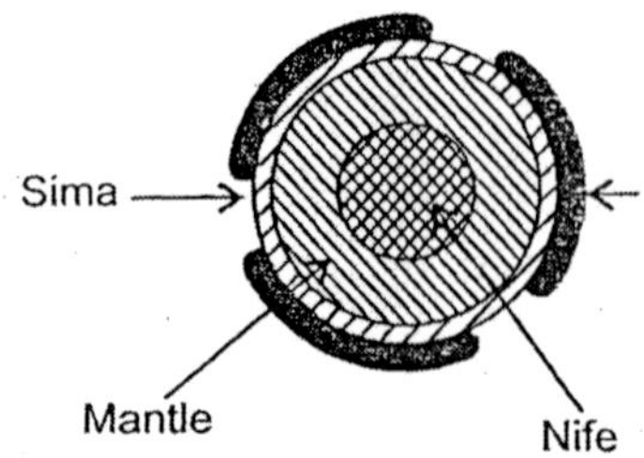

Fig. 3. Showing the structure of the earth (not to scale).

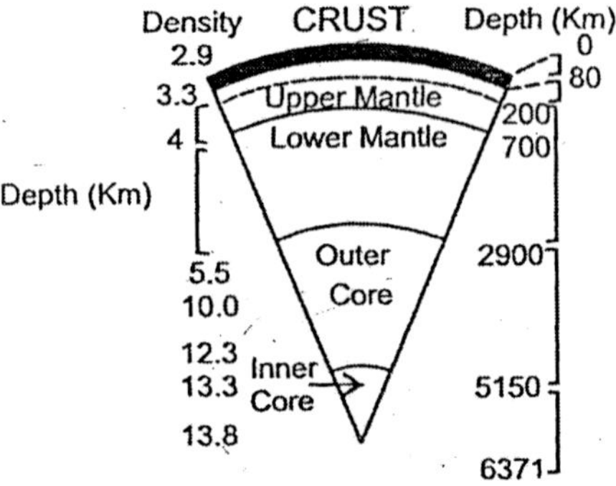

Fig. 4. Diagramatic presentation of different zones of the earth, their densities and their thickness.

The three zones or layers of the earth are as follows:

1. Lithosphere

The topmost layer of the earth or the rigid outer portion of the earth, is the lightest layer of the earth. In this layer the main earthquake waves travel at a nominal speed of 5 to 6 km per second, while transverse waves 2 to 3 km per second. The thickness of this layer is about 100 km with an average density of 3.5 grams cubic cm. The crust is thickest under the continents and reaches its thinnest point below the sea floor. It is mostly composed of granite rocks. The silica and aluminum are dominate constituents of this layer, that is why it is also known as sail (Si + Al.).

According to recent estimates the major constituents of the continental crust are silica or quartz (SiO_2) 62 percent, aluminum Oxide (Al_2O_3) 16 percent, iron oxides 6 percent, magnese oxides 3 percent, quicklime (CaO) 6 percent, sodium oxide (Na_2O) 3 percent and potassium oxide (K_2O) 3 percent.[16]

2. Pyrosphere

Below the top layer is the region of basalt rock, heavier than the first. Through this layer the main waves travel at a speed of 7.2 km per second where as transverse waves move at a speed of 4 km per second. It stretches for a thickness of 2780 km having an average density of 5.6 g. cubic cm. The upper portion of this layer is a mixture of sial and sima. Silica and magnesium are the dominant constituents of this layer, so it is also known as sial (Si + al). The mantle is believed to be separated from the crust by a thin zone called the Mohorovicic discontinuity or Moho. The matter of this layer is Wax like or elastic in nature. As a whole the mass of the mantle is 68 percent of that of the earth. This layer is the source of the energy and forces which are probably responsible for the phenomena of plate tectonics, orogeny, continental drift etc. Most basaltic eruptions (lava) are believed to have come from a depth of 100 to 120 km below the sea level from this zone.

3. Barysphere

Finally comes the central core of the earth. It is the heaviest layer composed mostly of iron and nickel, therefore also known as nife (Ni + Fe). The main waves (longitudinal waves) are unable to pass through this layer. The average density of this layer ranges from 8 to 11. This layer stretches from 2800 km upto the nucleus of the core (6370 km). The diameter of this zone is 6880 km. The temperature of the inner core may be as high as 6,700 degree F (3700 degree C).[20]

The presence of iron (ferrium) indicates the magnetic property of the earth's interior, besides the rigidity of the earth. Studies of seismic waves also indicate the division of the core into: a liquid outer core and a solid inner core.

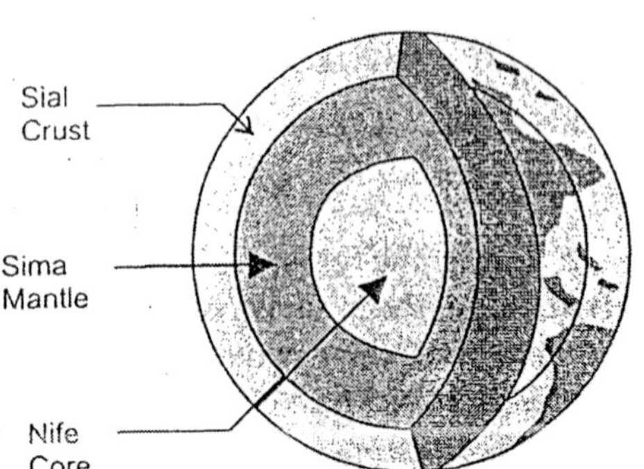

Fig. 5. Cross-Section of the Earth's Interior.

ORIGIN OF THE CONTINENTS AND OCEAN BASINS

The outer crust of the earth is divided into oceans and continents, both these relief features are of fundamental importance. The origin and nature of these have aroused as much controversy as the origin of the earth itself. The geologists say that the ocean basins are made up of heavier material than that of the material of continents, which is lighter.

About 70.8 percent of the total surface area of the world is represented by the oceans whereas only 29.2 percent being the continents. In other words the water covers an area of 141,05,000 sq. miles and the land surface has an area of only 55,500,000 sq. miles. Another characteristic is that the northern hemisphere consists of more land than water as compared to the southern hemisphere (water hemisphere).

If we study the globe, the following characteristic features of the distributional pattern of the continents and ocean basins would become clear.

(a) There is dominance of land areas (75%) in the northern hemisphere. Contrary to this water bodies dominate in southern hemisphere.

(b) Land and sea occupy an anti-podal position. Everywhere the land exists on the opposite side of the sea. On one end there is a land mass and opposite to it on the other end there will be a water body. For example, Europe and Africa are situated in the anti-podes of the Pacific Ocean, N. America to the Indian Ocean, Asia and Australia to Atlantic Ocean, while Antarctica to Arctic Ocean.

(c) All the continents (Americas, Europe, Africa, Asia and Australia) are almost triangular in shape, having their base in the north and tapering to the south. On the contrary, the chief water bodies of the world (Atlantic, Pacific, Indian Oceans) have their broad base in the south, while their apex towards the north, again having triangular shape.

(d) The great Pacific Ocean basin occupies almost one third of the entire surface area of the world. Another unique feature is that it is surrounded by young folded mountain systems, this arrangement being broken only for small stretches.

In deciding the problem, how continents and oceans originate, the chief considerations are:

(i) That the earth has cooled down from a gaseous state, hence its interior is still very hot.

(ii) That there is a definite order underlying the distribution of land and water on the surface of the earth. Taking these two factors into consideration various views, hypothesis and theories have been put forth by different geologists and scientists from time to time. We will examine here Wegner's and Lowthian Green's theories in detail, as they are mostly accepted, while the other concepts shall be briefly touched.

HISTORY OF DRIFT IDEA

As early as AD 1620, Francis Bacon drew attention towards the "Conformable Instances", in the configuration of the world. About 1800 AD Von Humboldt also suggested "the convexity of Brazil opposite the Gulf of Guinea" etc. Such occasional and indirect references to the idea of continental drift are found in the nineteenth century but Sinder's map (1858) showing South America in close contact with Africa was probably the first direct postulation of this theory.

F.B. Taylor in USA discussed continental displacements in 1908. Alfred Wegner put forward his theory of continental drift in 1915 in his German book (the origin of continents and oceans).

From 1920 to 1950, continental drift appeared to a few geologists and geophysicists as the obvious explanation of a wide range of phenomena but to most it seemed wrong - headed and impossible. [21,22]

The last heated controversy between the drifters and non drifters probably took place in 1948 in Birmingham. Subsequently, palaeomagnetic evidence and information about ocean floor brought about the revival of continental drift. But in it the jig-saw-fit of land masses, the geological resemblances, the palaeoclimatic features (direction of wind etc.), all play an important role and hence, by and large the evidences in favour of drift are conclusive.

TETRAHEDRAL CONCEPT OF LOWTHIAN GREEN

Tetrahedral Hypothesis of Lowthian Green is most significant of all the hypothesis based on geometrical principles. In 1875, Lowthian Green initiated and later on Fairbrain confirmed that the distribution of land and water on the surface has a tetrahedral arrangement. A tetrahedron is a solid body having four equal plane surfaces, each of which is an equilateral triangle. S.W. Wooldridge and R.S. Morgan (1959) also worked on the principles of Lowthian Green. They carried out experiments by crushing wrought iron tubes. L. Green based his theory on the following two basic principles of geometry.

(i) A sphere is that body which contains the largest volume with respect to its surface area.

(ii) A tetrahedron is that body which contains the least volume with respect to its surface area.

He opinioned that the crust of the sphere (earth) after cooling of its interior collapsed on the inner part and ultimately began to assume the shape of a tetrahedron.

Four oceans (the Pacific Ocean, the Atlantic Ocean, the Indian Ocean and Arctic Ocean) were created on the four plane faces of the terrestrial tetrahedron as they could retain water because they were lower than the level of the apices along which continents were formed.

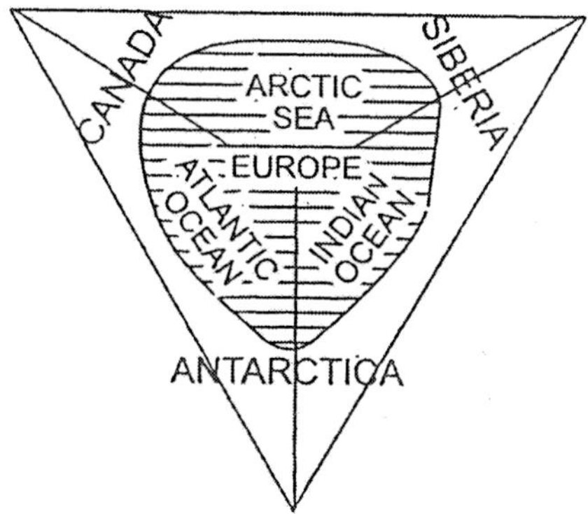

Fig. 6. Showing distribution of land and water on the tetrahedron.

Some Point in Support of the Tetrahedral Theory

1. Excess of land in northern hemisphere.
2. Almost all the oceans and continents are triangular in shape. The continents tapering towards south while oceans have their apex towards north.
3. All the landmasses are antipodal to water bodies e.g. Antarctica is antipodal to Arctic Ocean.

CRITICISM

The greatest drawback of this theory is that a tetrahedral earth cannot keep its balance while rotating. Secondly the earth is rotating so rapidly on its axis that the spherical earth cannot be converted into a tetrahedron while contracting or cooling. Thirdly, at present plate tectonic theory has validated the concept of continental drift, which goes against this theory.

WEGNER'S THEORY OF CONTINENTAL DRIFT

Prof. Alfred Wegener (1880—1930) of Germany who was primarily a meteorologist, propounded his concept of continental drift in 1912. It came before the common masses in 1922, when he elaborated his concept in his book "Die Entstehungder Kontinete and Ozeane", which was translated into English in 1924 entitled as origin of continents and oceans".

Following Edward Suess, Wegener believed in three layer system of earth i.e. sial, sima and nife. According to Wegener the mass of sial which constituted the condiments is floating on sima, a denser element. He assumed on the basis of fossils, vegetation and geological similarities that all the landmasses were united together and this composite mass was known as "pangaea", in carboniferous period. This single block of land was surrounded by a continuous ocean known as "Tethey's sea or Panthalasa", by Wegener. At some later stage due to the tidal drag by moon and the sun certain portions of this landmass drifted towards west and north. The result of westward drift was the formation of Americas while Atlantic Ocean came into existence in between Americas and Europe-Africa. On the same pattern, Australia drifted northwards and was separated from Antarctica. However, some scientists maintain that Antarctica and Australia drifted southwards separated from the mainland (Gondwanaland) and in its place came the Indian Ocean.

Wegener was primarily a meteorologist, in studying the climates of the world he came across the arrangement of jig-saw-fit of the continents of the world. He suggests the following instances in support of this arrangement.

He says the bulge of Brazil would fit in the Gulf of Guinea and the North American coastline would closely fit in the indentations of the coast of Scandinavia and western Europe. Similarly, the bulge of Ethiopia comes fit in the Gulf of Mexico and eastern coast of Africa may be fit in the curve of the coastline of western India and Pakistan. The northern tip of Australia can very well fit into the Bay of Bengal.

Wegener further resolved that with the westward drift the western edge of north and south Americas crumpled to form the mountain chain of Andes north Rockies south.

According to Wegener there is geographical similarity along both the coasts of the Atlantic Ocean. The rocks, fauna and flora are also the same on both the coasts. Geologically both the coasts of the Atlantic Ocean are identical. Du Toit, after detailed study of both rocks confirmed the Wegener's belief.

The behavior of lemmings (small sized animals) of Scandinavia to plunge into the sea also proves the fact that once these landmasses were united together.

It has been observed by actual calculations of the distance that Greenland is gradually drifting towards the north America at the rate of 20 feet per year, which proves that drift is possible.

The evidences of carboniferous glaciation or Great Ice Sheets has left their marks on southern part of America, India, Africa, Australia and Antarctica further prove the unification of all landmasses in a single block (Pangaea) during Carboniferous period.

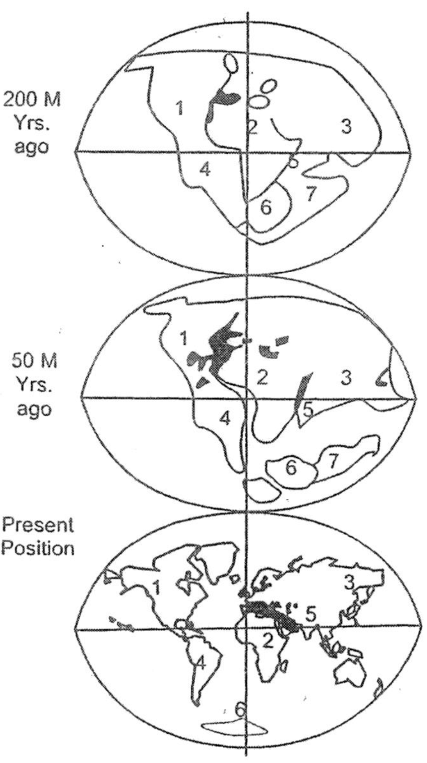

Fig. 7. 1. USA" 2. Africa" 3. China" 4. S. America" 5. India. 6. Antarictica 7. Australia.

CRITICISM OF THE THEORY

Wegner probably relied on the tidal drag by the moon and the sun as an important force for westward drift of continents. This meant that all continents had moved westward only. If this were so, then relatively some continents remained static or had a slower motion if tidal drag were to account for drift. It would have stopped the rotation of the earth altogether.

The theory presupposes sima to be liquid and sial to be solid, then how crumpling and folding was possible in solid sial (sedimentary rock).

CONTINENTAL DRIFT OR DISPLACEMENT THEORY OF TAYLOR

F.B. Taylor postulated his concept of horizontal displacement of the continents in 1908. He could not find any help from the contraction theory to explain the peculiar distribution of tertiary folded mountains, hence he gave his drift theory.

Taylor propounded this theory with the following assumptions.

According to him there were two landmasses: 'Lauratia', located in north pole and 'Gondwanaland', located near south pole. He further assumed that continents were made of sial (practically absent in the oceanic crust), which moved towards the equator. According to Taylor continents were displaced in two ways i.e. equatorward and westward. But the main driving force in both cases was tidal force of moon. This movement resulted into tentional force near the north pole which caused stretching, splitting and rupture in the landmasses. Consequently, Baffin Bay, Labrador Sea and Davis Strait were formed.

Similarly Gondwanaland was split into several parts, giving birth to Australia, Ross Sea, Arctic Sea, Greenland and Siberia. The Himalayas and Alps were formed with the equatorward drift of both landmasses from poles while the Rockies and Andes were formed due to westward movement of the Landmasses.

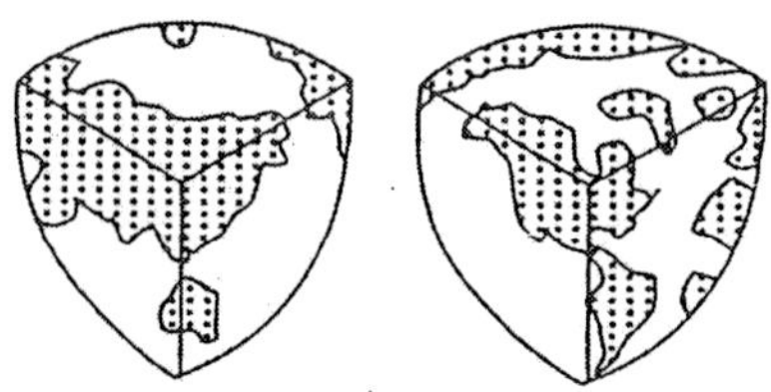

Fig. 8. Showing Tetrahedal Earth (After Griffith Taylor).

CRITICISM

1. Since F.B. Taylor's main concern was to explain the origin of the tertiary folded mountains, hence he made the continents to move thousands of kms from poles to equator, whereas 32 to 64 km drift would have been sufficient enough for the purpose.
2. The mode of drift has also been found erroneous. If the tidal force was so enormous that it could displace the landmasses to thousands of km apart then it might have stopped rotation of the earth. A. Holmes says that neither tidal nor external force can drift the continents but this kind of force must come from within the earth. Holmes has rightly remarked, but Taylor must be given credit for making an independent and slightly an earlier start in this precarious filed.

KELVIS HYPOTHESIS

This was the first hypothesis about the origin of oceans and continents. He says the continents originated as nuclear dots in the gaseous phase of the earth. But this theory has been criticized on the grounds that it does not explain the exact location of the continents. The dots must have been formed at random while the continents have an antipodal arrangement. Secondly it does not explains the difference in material of continents and oceans.

LOVE'S THEORY ABOUT THE ORIGIN OF CONTINENTS AND OCEANS

Prof. Love resolved the following facts about the origin of continents and oceans:

(a) The centre of gravity does not coincide with geometric centre of the earth. It depends upon the mass of the material. Hence it shifts towards the area containing heavier material.

(b) There is a lack of coincidence between the centre of the gravity and the geometric centre. This lack of co-incidence generate forces. These forces deform the earth's body so that it can gain stability and equilibrium.

(c) The deformation resulted in the creation of continents and oceans. The protrusions became the continents, while the oceans took their place in sagged parts of the crust.

There are so many flaws and draw backs in this theory, on account of which it is not acceptable by masses.

E.SUESS'S THEORY ABOUT THE CONTINENTS AND OCEANS

E. Suess assumed that the present continents made of resistant hard materials were original shields or landmasses. The present non-resistant parts were formerly all covered with water. In the intervals between various earth movements, the old masses sank and extended the extent of the water bodies while in the area of non-resistant parts of ocean beds became folded and formed land. The rigid masses of the earth's surface are as follows:

1. The Baltic Shield

The continent of Europe has grown up around the Baltic Shield, the rock is exposed near Baltic Sea.

2. The China Tableland

The Chinese Tableland which is now covered by newer depositional beds.

3. The Siberian Shield

It is also known as Angaraland, extends in the west upto Ural mountains and upto Byrranga Hills in the Taimyr Peninsula in the north.

4. Deccan Plateau

The Indian Tableland, bounded by Indian Ocean on the east and west is said to have formed a part of Gondwanaland.

5. Canadian or Laurentian Shield

It is bounded on west by Rockies on the north by Arctic Oceans and on the south and east by lakes and the Labrador. Colorado Plateau is also an integral part of it, which lies in its south west.

6. Brazil-Guiana

Rigid mass is the nucleus of the south American continent.

7. Antarctica

Antarctica is itself a rigid mass although not a great deal is known about this region.

8. Africa

Africa is one single rigid mass especially south of the Congo.

All the above natural features are formed of old hard rocks and present a massive structure. Close to and around these have been belts or zones which were mobile during the long geological history. These may be called Geosynclinal areas wherein were formed the great mountain ranges of the world. Mediterranean Sea is the remnant of the former sea of Tetheys, being a very solid proof and a real supporting point of this theory.

CONVENTION CURRENT THEORY OF PROF. A HOLMES

Prof. Holmes reached the conclusion that the earth is cooling through convection. Hot sima goes up in currents while the currents of colder elements comes down or descends. The substratum is both rigid and viscous. But the convection currents are not checked by this viscosity. On the other hand, the portions of the continents sunk or embedded in the sima layer or basalt do obstruct the movement of these currents. The result is that these currents seek their way by pushing through and due to this push the continents begin to drift.

CHAMBERLAIN AND MOULTON'S HYPOTHESIS

Chamberlain and Moulton both together based their theory on the planetesimal hypothesis of origin of the earth. The whole solar system came out a very finely divided solid particles which became compact after maturity into planets and satellites. Earth's initial view was due to an unequal in fall of planetesimal material. The excess of the material fall resulted in continents and other became oceans.

CRITICISM

(a) The random in fall of planetesimal material cannot display an antipodal arrangement.

(b) If the infall is continuous the irregularities should also be continuous. But oceans and continents once formed contemporaneously have by and large remained unchanged.

JOLLY'S THEORY OF THERMAL CYCLES OR RADIO ACTIVITY

Jolly works on the footprints of Wegener and thinks that the continents are sheet like masses of sial floating on a denser element sima. Both of these are supposed to be radio-active, on account of which it is also known as Radio-Activity Theory. According to Jolly there were large vertical movements of the land masses, due to the variation in the density of the sima. Jolly assumes the following points about his theory.

1. The continents have a vertical movement not horizontal drift.
2. The sima has sufficient amount of radio-active element which produces heat on disintegration.
3. With sufficient heat sima melts and its density is reduced, with the result the sial sinks into it.
4. With successive submergence of the continents, the accumulated heat is dissipated, sima again becomes solid and the result is that the continents are uplifted.

Criticism

1. Why does the ocean bed also not melt with sima? Jolly maintains that the ocean bed remains intact but gives no reason for it.

2. Prof. Holmes say radio-activity impossible.

Sollas's Hypothesis

Sollas assumption is based on following basic facts.

(a) The earth was originally in a molten state.

(b) It could be pressed at one place to produce bulge at other.

(c) These hollows and bulges are produced by unequal pressure acting against the then existing molten conditions.

(d) The depressions became oceans and bulges became continents.

Criticism

Lothean Green has proved antipodal system of the arrangement of continents and oceans, whereas Sollas has maintained random system in the arrangement.

The original differences in the atmospheric pressure could not have produced so large and deep oceanic basins separated by so large continental areas.

4
Earth Movements

The landforms on the surface of the earth are never permanent in nature but tend to change. Land rises and sinks; points on the globe move closer together or further apart; island appear or disappear. While the land is seen tilting upwards in one part of the world and gradually sinking under the sea at another point. The land moves up and down as readily as sideways, with the result that distances are altered; new island; are born while old ones vanish; Oceans juggle with continents and if beaches are washed away at one place, thousand, of miles are left day at another place by shrinking sea.

The earth is slowly but continuously changing its face due to pressure, movement, tension, compression and fracture forces of the earth. The mountains and plateaus are levelled down to plains while the green plains and ocean beds may be uplifted to form highlands.

Thus the forces affecting the crust of the earth or geological processes is of paramount significance. In fact change is the law of nature.

The geological changes are generally of two types.

(a) Long Period Changes

Long period Changes occur so slow that man is unable to notice such changes because of his short span of life. These changes are brought about mainly by erosion and denudation through different agents, such as running water, glacier, winds and underground water.

(b) Short Period Changes

On the contrary, the short period changes take place so suddenly that these are noticed with in few seconds to few hours. These changes are brought about by seismic events, volcanic eruptions etc.

On the basis of the sources of the origin the forces, which effect the crust can be divided into two broad categories i.e. the endogenic forces which act from within the body of the earth and exygenic forces which react from above.

1. EXOGENETIC FORCES

Exogenetic forces refers to the external forces of denudation or destructional forces are always originated from the atmosphere. These forces are continuously engaged in changing the face of the earth. The exogenous processes are the changes in landform brought by external agents. River, wind, glacier and sea waves are the chief external agents, which change the land form through the processes of gradation, degradation, aggravation and weathering. Thus, their work is threefold viz. erosion, transportation and deposition. Their work takes place in atmosphere and hydrosphere. They act slowly, we ar down the high lands and deposits materials in the low lands.

2. ENDOGENETIC FORCES

The forces coming from within the earth or taking place in the interior of the earth are called endogenetic forces. They work abruptly as well as slowly. They produce movements on the earth's crust and often lead to either mountain building, volcanic eruption or earthquakes. Hence, they are also known as earth movements. The endogenetic forces cause two types of movements in the earth viz. horizontal and vertical movements. These movements give birth to numerous varieties of relief features on the earth's surface e.g. mountains, plateaus, plains, lakes, faults and folds etc. The origin of endogenetic forces is related to thermal conditions of the interior of the earth. The horizontal and vertical movements are caused due to contraction and expansion of rocks on account of varying thermal conditions and temperature changes inside the earth. On the basis of the intensity the endogenetic forces are divided into two major divisions as under:

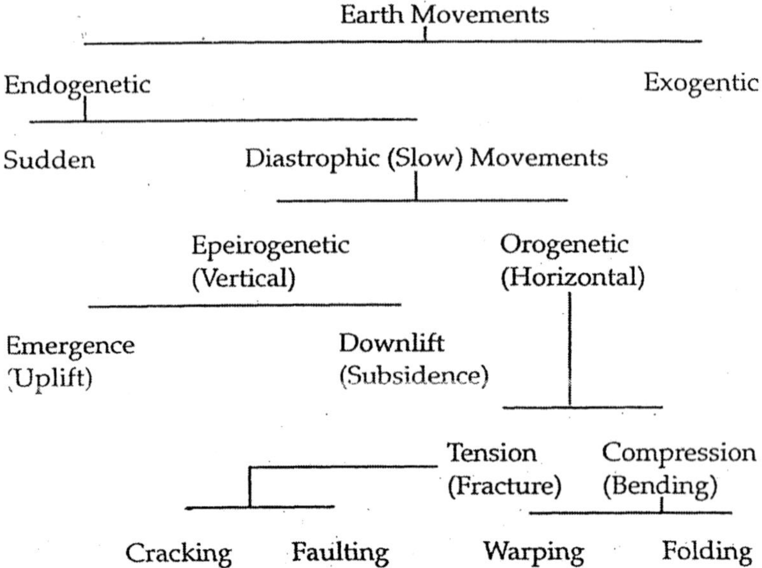

a) Sudden Forces or Movements

Earthquakes and volcanic eruptions (seismic events) are called as sudden movements and the forces responsible for their origin are known as sudden forces. These extreme events become disastrous hazards when they occur in densely populated localities, as these forces work very quickly and their results are seen within minutes. These forces create certain relief features, such as Deccan plateau of India, Columbia plateau of USA, lava plains, faults, fractures and lakes. Geologically these forces are termed as "constructive forces".[25]

b) Diastrophic Forces or Movements

Diastrophic forces include both vertical and horizontal movements which are caused due to forces deep within the earth. The operation of these forces is in a gradual process and show their effects after thousands and millions of years. They bring out and produce meso-level relief features, like mountains, plateaus, plains, lakes, folds and faults etc. These movements are further sub-divided into two groups.

(i) Epeirogenetic Movements

It is a combination of two words epiros meaning continent and genesis meaning origin. This movement causes upliftment and subsidence of the continental masses through upward and downward movements respectively. These movements do not disturb the horizontal arrangements of the crust. The best example of subsidence is that of land towards the west of the Western Ghats. The elevation of certain coastal areas of Florida, Sweden and West Indies are the best example of emergence. The entire Scandinavian peninsula is gradually submerging while its northern part is emerging.

(ii) Orogenetic Movements

The word has been derived from Greek word oros, meaning mountain and genesis meaning origin, thus the movements are related to the orogeny or the mountain building process. These forces work horizontally and cause crumpling, faulting, folding etc. in the crust of the earth. The orogenetic forces are sub-divided into two divisions according to their work, which are as under:

(a) Tensional Force

The tensional movement or force which operates in opposite direction is also called as divergent force or movement. These forces create rupture, cracks, fracture and faults in the crustal parts of the earth.

(b) Compressional Force

The compressional force or movement operates face to face and is also called as convergent force. This force causes crustal bending leading to the formation of folds or crustal warping, rise or subsidence of crustal parts.

STRUCTURES RESULTING FROM EARTH MOVEMENTS

The different relief features of the earth are in fact born out of constant struggle between endogenetic forces and exogenetic forces or movements contraction and expansion the two significant forces are very much related to one another. They are also called as compression and tension. Compression in one region of the earth's crust is bound to produce tension at another place and vice versa. While the forces of compression produce folds and bends in the crust of the earth, the forces of tension are responsible for giving rise to faults and joints. These relief features or structures are discussed below with more detail one by one.

(I) Crustal Bending

The crustal rocks are bent due to the face to face horizontal force or compressional movement which is known as crustal bending. The crustal rocks undergo the process of crustal bending in two ways e.g., warping and folding. The crustal parts are either raised upward or downward, known as up warping or down warping. When the compressive forces cause buckling and squeezing of crustal rocks, the resultant mechanism is called as folding.

FOLDING

The deformation or bending of rocks strata produced by compressional forces is termed as folding. When the rock strata are compressed from side to side, a part of the crust is raised into folds. Folding occurs when horizontal forces act towards a common point or plane from opposite directions. Rock folds look like waves of water on a lake surface, resulting in a series of troughs and crests. The crests are the up folds or 'anticlines', while the troughs are the downfold or 'synclines'. The force of compression leads to contraction of folds while the force of tension leads to expansion of the folds. Due to folding, a ridge and the valley topography is formed. Folding has caused most of the world's great mountain systems. Rockies in North America and Andes in South America are the typical examples of folds mountains, besides the Alps mountain system.

Depending on the intensity of the movement, force or push and the rocks structure, different classifications of the folds have come into existence. These are symmetrical, asymmetrical, overturned, monoclinal, Isoclinal, homoclinal, recumbent, fan fold, open, closed and napper folds. All these terms would be treated one by one in detail under separate topic viz. 'Mountain Building', in the same chapter.

(II) Crustal Fracture

Crustal fracture refers to displacement of rocks along a plane due to tensional and compressional forces. It depends on the intensity of the tensional forces and the nature or strength of rocks. The crustal rock sometimes only cracks, when the tension is moderate while it is subjected to dislocation and displacement (faults), when the rocks are subjected to intense tensional force. The crustal fracture is generally divided into two categories i.e. faults and joints. A fracture becomes fault when there is a considerable displacement of rocks on both sides of a fracture and parallel to it a joint is defined as a fracture in the crustal rocks.

FAULTING

Faulting refers to a fracture in the earth's crust along which movement has taken place and where the rock strata on the two sides do not match. Although the movement is generally vertical, a fault may take place in any direction. The compression produces folded mountains, where as tensional forces produce faults and joints. The tensional force is of two types local and regional. The local tension may be due to intense compression and produce over thrust fracture. The regional tension, however, alternates with compression and produces faults.

The change of level on one side as compared to the other is known as the "throw", differs from less than a centimeter to hundreds of meters. The amount of lateral displacement is called the "heave". When the crust is broken, on one side of the break the strata drops in relation to other. This is known as fault and the side of fault where the strata have moved relatively downwards is called the "downthrow" side, the other being the "upthrow" side. The normal fault is one where in the "fault plane" is vertical or is so inclined that the downthrow is one the dip side of the fault plane. If the break in the crust is distributed over a number of planes generally parallel to one another, it is known as a "fault zone".

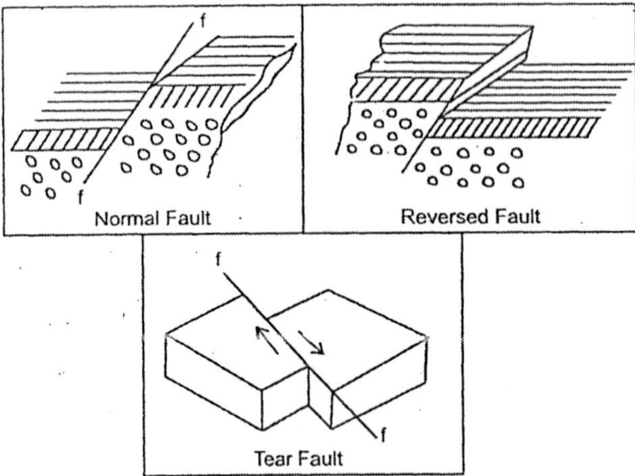

Fig. 1. Fault.

TYPES OF FAULTING

The fault may result in single form or in manifold pattern. The manifold faults may either be arranged in a pattern like the steps of a ladder. The number of parallel faults are thrown down in the same direction the feature formed is known as "step faulting". On the other hand, when the parallel faults are arranged in pairs, important relief features are formed. If the block of crust between two parallel faults had been downthrown, a rift valley is formed. On the contrary if the block has been upthrown or raised a "horst" or 'block mountain', comes into existence.

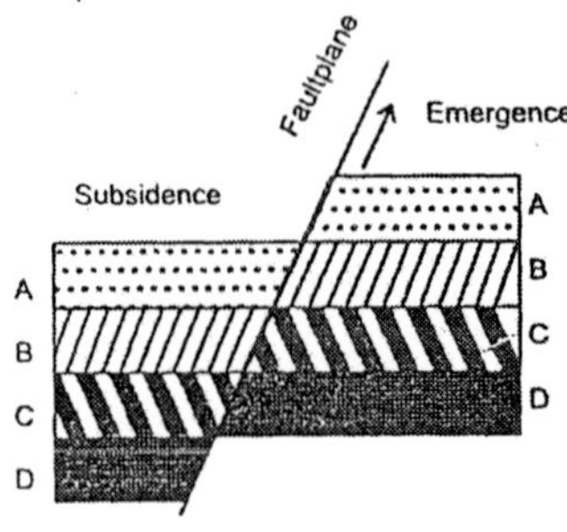

Fig. 2. Tear Fault

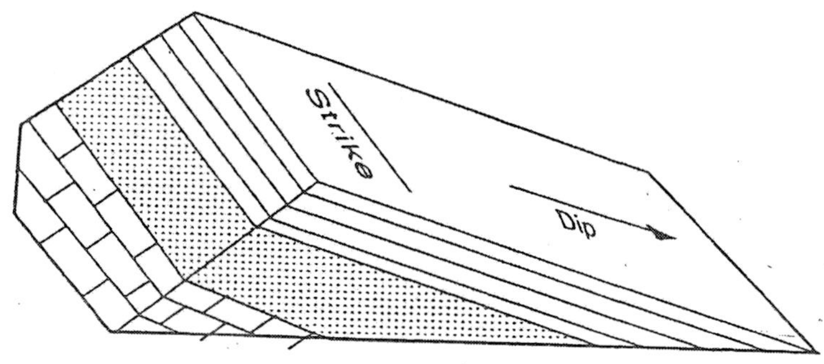

Fig. 3. Rock strata illustrating Dip and Strike

(a) Normal Fault

Normal faults are formed due to the displacement of both the rock blocks in opposite directions due to fracture consequent upon greatest stress.

(b) Reversed Fault or Thrust Fault

A fault in which the upper rock strata have been pushed forward over the lower strata, the upthrow being on the dip side of the fault plane. It may be formed by the breaking of an over turned fold, or over fold, at the sharply bent crest of the anticline.

(c) Lateral or Tear Fault or Strike Slip Fault

Lateral fault is the fault in which the movement of rocks is mainly horizontal rather than vertical. The terms regarded as synonymous with lateral fault are tear fault, wrench fault and strike-slip fault. Great Glen of Scotland in Britain with a gap of 65 miles is the typical example of tear fault.

(d) Step Fault

When a series of faults occur in any area in such a way that the slopes of all the fault planes of all the faults are in the same direction, the resultant faults are called as step faults.

(e) Oblique fault or Oblique Slip Fault

Oblique fault refers to a fault whose strike is oblique to the strike of the beds it traverses.

(f) Hinge Fault

In hinge fault the displacement increases from zero to a maximum along the strike.

(g) Point Fault

In point fault one block appears to have rotated about a point on the fault plane. It is also called scissor fault.

(h) Rift Valley or Graben or Trough Fault

A valley which has been formed by the sinking of a portion of the crust of the earth between two roughly parallel faults is known as rift valley. Such a valley is often long in proportion to its width. The sides of this valley are high and steep. The term graven derived from German word is also regarded as synonymous with rift valley. The Dead Sea of Israel, the Red Sea, the lake of Africa and the Rhine valley of Germany are the important examples of Rift valley.

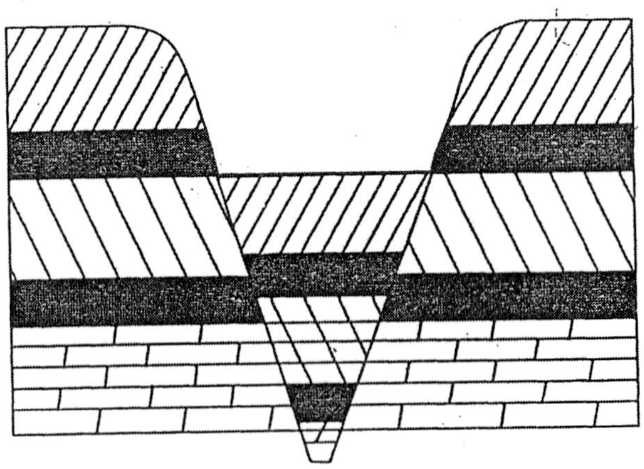

Fig. 4. Rift Vally

(i) Horst or Block Mountain

Horst, a German term, is an elevated block of rock between parallel faults which has reached its position either through uplift between the faults or through the sinking of the beds outside the fault.

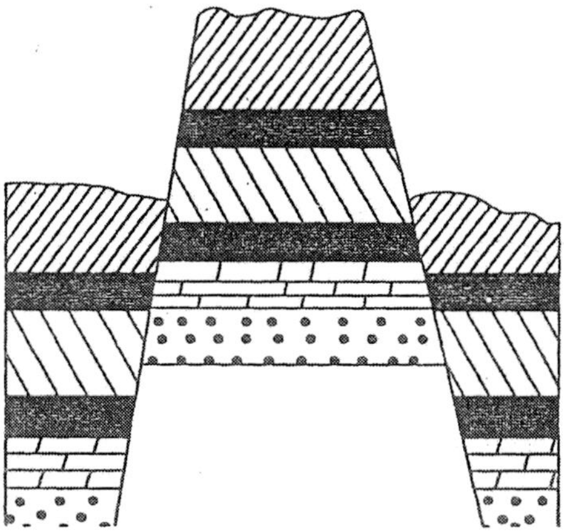

Fig. 5. Block Mt. or Horst

JOINTING

Joints are those crustal fractures in which no upthrow or downthrow has taken place. It is a crack in the crust formed along a plane of weakness or joint plane, .with a little or no movement between the blocks. The joints are usually formed by local tension e.g. due to drying up, the sedimentary rocks contract and are subject to general tension, giving rise to joints. Secondly, by cooling and consolidation of igneous rocks tension is produced by contraction, resulting in a complex system of joint. Thirdly, due to crystallisation of certain rocks, splitting planes occur and develop into joints. Lastly due to twisting of anticlines and synclines the joints are formed in the crust of the earth. These joints may be wide in case of anticlines, while narrow in synclines.

TYPES OF JOINTS

There are various types of joints formed according to the rock material, type and intensity of force or tension. Some important types of joints as follows:

(a) Column Like Joints

Column like joints are formed by the cooling of the molten igneous rocks.

(b) Cleavage Planes

Cleavage is the splitting of a rock, that has been put to great pressure, into thin sheets or slabs. The direction of cleavage is often parallel to the stratification of the rocks, but generally inclined at right angles to it. Jointing is due to planes of weakness in the rock. As in case of roofing slate, cleavage planes are well developed, parallel to the bed or strata in sedimentary rocks.[26]

(c) Master Joints

These joints are most dominant in any type of rock and are known as master joints. These consists of two parallel sets e.g. dip joints, parallel to the direction of the greatest slope and strike joint, parallel to the line drawn at right angles to the dip. Because of these master joints, the pattern of different rocks is very much complicated. The igneous rocks break into polygonal column while the sedimentary rocks break into cuboidal blocks.

The joints are usually the planes of weakness in a rock. Whether they are open or laminated they exercise a very great control on the details of landscape formation.

ISOSTASY

Isostasy is the theory or concept which supposes that the adjoining regions of the crust of the earth are in balance among themselves. Different relief features of the earth (mountains, Plateau, Plains and Oceans) are probably balanced by certain definite principle, otherwise these would have not been maintained in their present form. The violent earth movements and tectonic events start whenever this balance is disturbed.

In the words of Holmes, the term isostasy (Greek word isostasios in equipoise - equal balance) was first proposed by the American geologist C.E. Dutton in 1889, "for the ideal condition of gravitational equilibrium that controls the height of continents and ocean floors in accordance with the densities of their underlying rocks".[23]

According to J.A. Steers (1961), the above doctrine states that wherever equilibrium exists on the earth's surface, equal mass must underlie equal surface area

In simple words isostasy is the name given to all those complex forces which keep to all those complex forces which keep the varied features of the earth in one balance or equilibrium. This idea of balance led to the belief that 'sial' which constituted the light continents were floating on 'sima', a denser element. C.E. Dutton assumed that every continent displaces from below it the liquid basalt equal to its weight. Thus to a very great depth below the apparent height of the continents there are only lighter materials. On the contrary the ocean bed is resting on heavy basalt. Hence in the ocean beds there is a very thick heavy layer which is the main reason of the balance between the continents and the oceans. This further confirms the fact that the comparative weight of the landmasses and water bodies is the same. But in order to determine this, one has to understand the gravity of the earth.

The longitude, relief and height above sea level tend to increase or decrease the gravitational pull of the earth. But the actual observations show that the calculated values of gravity are different from the observed value. In order to strike an agreement between the two, Hayford and Bowie imagined that below the earth's surface, there is a certain level beneath which the densities were the same throughout. To this they named as "compensation level", above which the variations of density do exist. In order to explain this difference they suggested that if adjoining columns of equal cross-sections are set on this compensation level, the density would remain to their height in an inverse proportion so that they exert equal pressure downwards.

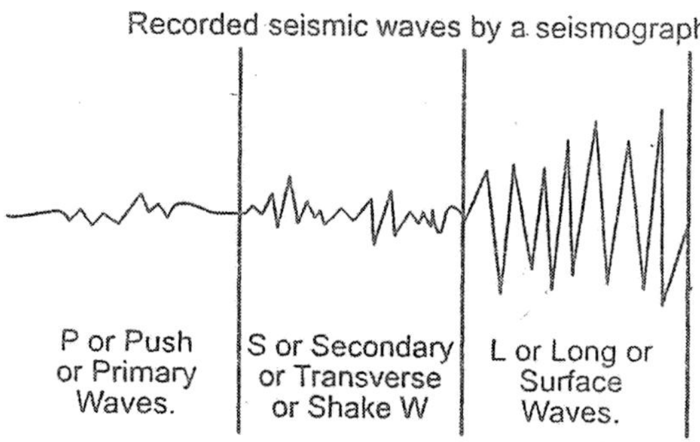

Fig. 6. Recorded seismic waves by a seismograph.

It means that the different regions of different density are quite disconnected from place to place. Whenever, due to the deposition of eroded material the pressure increases on a particular region, so much that the isostatic balance is disturbed, the higher regions, which have been eroded, grow still higher, while the depositional regions sink down as the magma flows towards the higher regions and the balance is again restored.

The isostasy is a hypothesis and a fact which is now widely believed in and explains to a very large extent the present distribution of continents of oceans on the surface of the earth. When a part of the earth becomes overloaded with a large amount of sediment as happens even now in deltas of large rivers, the crust of the earth sinks under the weight. The tilted shore-lines of Great Lakes of North America and moving upward of the Scandinavian Peninsula at the rate of 3 feet a century represents the isostatic recovery of earth's crust from such weights. Another example of such evidence is that Alaskan shoreline is tilting upward and the east coast of Africa is rising, while the eastern coast of North America is gradually sinking in the sea.[24]

THE CONCEPT OF SIR GEORGE AIRY

According to Airy the inner part of the weight of the mountains cannot be hollow, rather the excess weight of the mountains is compensated or balanced by lighter material, below. He assumes that the light sial is floating on heavier sima. Hence Himalayas are floating in denser glassy magma. Airy further assumed that Himalayas are floating in the denser magma with their maximum portion sunk in the magma in the same way as a boat floats in water with its maximum part sunk in the water. Infact, this concept involves the principle of flotation. If we assume the average density of the crust as 2.67 and that of substratum as 3.0, so, for every one part of the crust to remain above the substratum, nine parts of the crust must be in the substratum according to the principle of flotation. He further claims that density zones not change with depth, that is, "uniform density with varying thickness".

Airy proved his concept by making an experiment. He took several pieces of iron of varying lengths and put them into a basin full of mercury. These pieces sunk upto varying depths depending on their lengths. He demonstrated the same experiment by wooden pieces by putting them into the water basin.

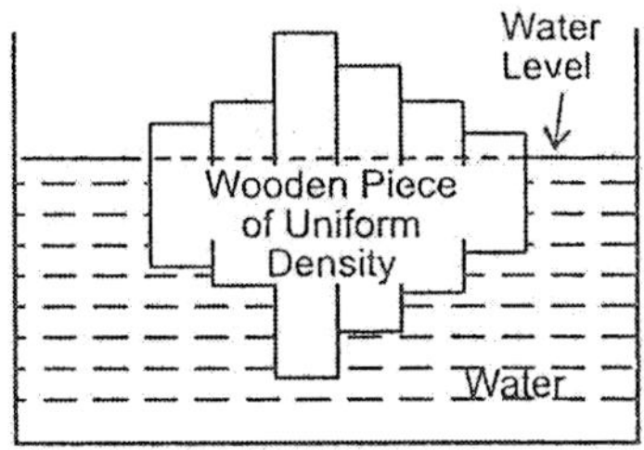

Fig. 7. The Concept of Sir George Airy.

CRITICISM OF THE THEORY

If we accept the Airy's views of isostasy, then Himalaya's roots would reach a great depth of 79634 meters. On reaching such a depth they would melt due to a very high temperature as it increases with increasing depth, at the rate of one degree C per 32m.

THE CONCEPT OF ARCHDEACON PRATT

The density of mountains is less than the density of plateau, that of plateau is less than the density of plain and the density of plains is less than the oceanic floor and so on. It means that there is inverse relationship between the height of the reliefs and density.

According to Pratt there is a level of compensation above which there is variation in the density of different columns of land but there is no change in density below this level. Density does not change within one column but it changes from one column to other columns above the level of compensation. Thus the central theme of the concept of Pratt on isostary may be expressed as, "uniform depth with varying density".

The Pratt's concept of inverse relationship between the height of different columns and their respective densities may be expressed in the following manner - "bigger the column lesser the density and smaller the column greater the density'. According to Pratt density varies only in the lithosphere, thus his concept of isostasy was related to the ' Law of compensation" and not to the law of flotation.

While making a comparative analysis of the views of Airy and Pratt on isostasy, Bowie has observed that, "The fundamental difference between Airy's and Pratt's views is that the former postulated a uniform density with varying thickness and the latter a uniform depth with varying density."

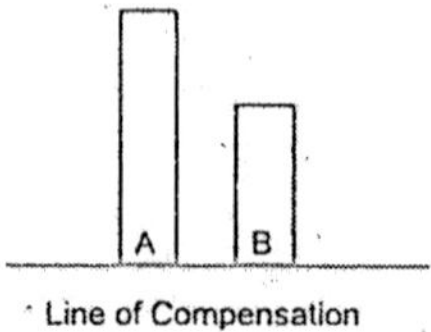

Fig. 8. Showing the line of compensation according to Archdeacon Pratt.

THE CONCEPT OF HAYFORD AND BOWIE

Hayford and Bowie propounded their concept of isostasy on the basis of Pratt's view. According to them there is a plane where there is complete compensation of the crustal parts. The densities vary with elevations of columns of crustal parts above this plane of compensation. The density of the mountains is less than the ocean floor according to them. In other words the crust is composed of lighter material under the mountains than under the floor of the oceans. There is such a zone below the plane of compensation where density is uniform in lateral direction. Thus according to Hayford and Bowie there is inverse relationship between the height of columns of the crust and their respective densities (as assumed by Pratt) above the line of compensation. The columns having the rocks of lesser density stand higher than the columns having the rocks of higher density.

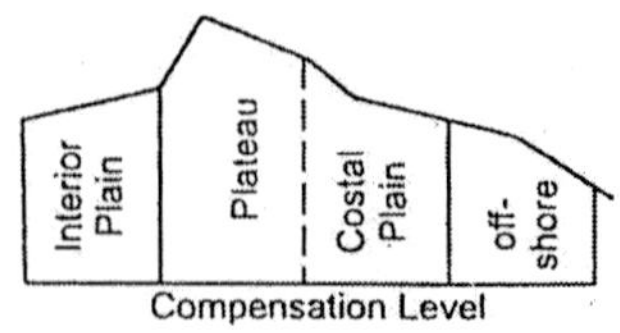

Fig. 9. Showing the explanation of views of Hayford & Bowie on isostasy.

THE CONCEPT OF JOLY

Joly presented his views about isostasy in 1925. He contradicted and disapproved the concept of Hayford and Bowie about the existence of level of compensation at the depth of about 100 km on the grounds that the temperature at this depth would be so high that it would melt everything.

According to Jolly there exists a layer of 16 km thickness below a shell of uniform density. The density varies in this zone. It thus, appears that Jolly assumed the level of compensation as not a linear phenomenon but a zonal phenomenon. In other words, he did not believe in a 'Level of Compensation', rather he believed in a zone of compensation.

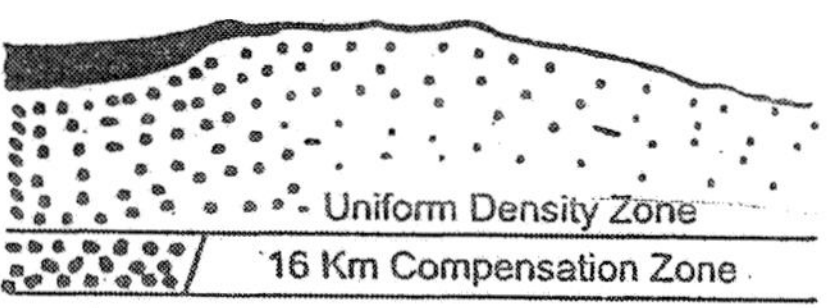

Fig. 10. Finer dots indicate lighter materials while larger dots represent denser materials.

THE CONCEPT OF PROF. HOLMES

Prof. Arthur Holmes has followed the views of Airy saying the upstanding crustal parts are made of lighter material and in order to balance them major portions of these higher columns are submerged in greater depth of lighter material. According to Holmes the higher columns are standing because of the fact that there is lighter material below them for greater depth, whereas there is lighter material below the smaller columns upto lesser depth.

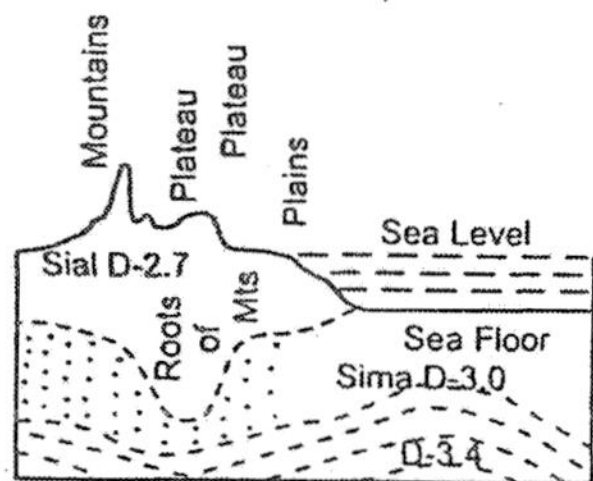

Fig. 11. Isostasy based on A. Holmes and D.L. Holmes, 1978.

EARTHQUAKES

An earthquake is a vibration or jerking motion in the rocky shell of the earth's crust backwards and forwards. As water of a pond generates waves in all directions when struck by a stone, so the crust is set into tremors or waves by a sudden blow from external and internal causes. These causes are folding, faulting and slipping of rocks. In the mountainous areas the landslips or the fall of avalanches often cause light shocks of minor intensity. Violent eruptions of lava and falling of underground caves often cause vibrations on the earth's crust. In other words, and earthquake is a major demonstration of the power of the tectonic forces caused by endogenetic thermal conditions of the interior of the earth.

The intensity or magnitude of energy released by an earthquake is measured by the Richter Scale devised by Charles F. Richter in 1935. The number (M) of the scale ranges between zero to 9. The shock produced is sometimes highly disastrous to human life and property. The greatest earthquake of the world ever recorded are 8.4 Bihar in 1935 and 8.6 Alaska (USA) in 1964.

The point of origin of an earthquake is called as hypocenter or focus. Its depth varies from 30 km to 700 km below the ground (Bihar - Nepal earthquake 1988). The place immediately above the focus on ground surface is known as epicenter. The waves generated by an earthquake are called seismic waves, which are recorded by an instrument called seismograph. The science of earthquake is called seismology which has come from the Greek words seismos meaning an earthquake and logos meaning science.

CAUSES AND TYPES OF EARTHQUAKES

Aristotle believed that underground air in its endeavour to come out shakes the earth. Lucretious believed that the collapse of underground caves produce shocks. Magnati in 1688 A.D. proposed the action of wind, water and heat and the concentration of coaly-matter for the cause of earthquakes. It is due to the research of Mallet, Milne, Reid, Imamura, Omori and others that we have at present a fair knowledge of earthquakes though there are many points yet to be made clear.

According to the causes, the earthquakes may be generally divided into the following three groups.

(I) Volcanic Earthquakes

Volcanic errupations and explosions can produce severe earthquakes, so it is considered to be one of the major causes. In fact, volcanicity and seismic events are so intimately related to each other that they become cause and effect for each other. The earth tremors are major precursor events of possible volcanic eruption in immediate future in any region. The explosive violent gases during eruption try to escape upward and hence they push the crustal surface with great force from below causing tremors of high magnitude. Whenever these gases become successful in breaking weak crustal surface, the magnitude of earthquakes depends upon the intensity of volcanic eruption. These earthquakes are confined to volcanic areas. Examples, severe earthquakes caused by violent explosions of Krakatau (Java and Sumtra) volcano in 1883 and Etna Volcano in 1968 which killed about 40,000 people. Practically, they produce no geological effect.

(II) Tectonic Earthquakes

The word tectonic has come from Greek word tekton meaning a builder, referring to the changes brought about in the earth's crust. The earthquakes are mainly caused by tectonic occurrences like faulting, folding, breaking, raising and sinking of the crust, or vapour seeking to escape from it. The endogenetic forces cause isostatic disequilibrium in the crustal rocks which results in the shaking of the earth. Infact, sudden dislocation of rock blocks caused by both tension and compressive forces cause immediate tremors due to sudden maladjustment of these blocks. The Assam earthquake of 1950, Bihar earthquake 1934 and Koyna (Maharashtra) earthquake 1967 are the examples of tectonic cause. Commonly the displacement is too deep to leave any visible sign on the surface, like that of San Francisco (California) earthquake of 1906.

ELASTIC REBOUND HYPOTHESIS

Prof. H.F. Reid after a careful study of the San Francisco earthquake disaster of 18th of April, 1906 advanced his Elastic Rebound Hypothesis about the origin of the earthquake. According to him stresses on the two sides of the fault were accumulating and produced bending of rocks as they are elastic in nature. But when the rocks could bear no more strain the breaking with sudden displacement of the elastic rocks on the two sides of the fault took place. It produced a blow to the upper rocks on one side of the fault plane and to the lower rocks on the other side (See Fig. Elastic Rebound).

The California earthquake was caused due to the formation of 640 km long San Andreas and Sagami fault. Japan Bay earthquake (1923) was also caused by a fault.

PLATE TECTONIC THEORY

Recently, a new theory has been propounded named as Plate Tectonic Theory, which is accepted as most possible explanation of the causes of earthquakes. According to it, the crust is composed of solid and moving plates having either continental crust or oceanic crust or both. The earth's crust consists of 6 major plates and 20 minor plates. These plates are constantly moving in relation to each other due to thermal convective currents, originating deep within the earth. Thus, all the tectonic events take place along the boundaries of these moving plates.

(III) Surface or Man-Induced Earthquakes

The earthquakes may be generated by both natural process i.e. landslide, slips on steep coasts, avalanches and underground roof collapse etc. or caused by human activities such as pumping of water or mineral oil from the earth, construction of dams, reservoirs and roads etc. Examples, 1931 earthquake of Greece due to Marathon Dam 1936, E.Q. of Hoover Dam (USA) due to Lake Mead, 1967 Koyna E.Q. (Maharashtra) due to Koyna reservoir, 1911 Turkistan E.Q. due to huge landslide in Pamir. Slight tremors are often recorded in Alipore (Calcutta) observatory which are caused by the dashing of the sea waves against the Coromodal Coast.

SUBMARINE EARTHQUAKES

Earthquakes often occur under the sea water on the beds of the seas. They cannot spread very far in water, but passing ships are often effected by them. Vertical displacement on the sea bottom causes huge sea waves producing devastating effects on the sea coasts. Examples, 1703 and 1896 damaged the Japanese coast and 1735 Lisbon earthquake.

DISTRIBUTION OF EARTHQUAKES

If we look at the world distribution map of earthquakes it appears that most of the world earthquakes occur in the zones of young folded mountains, faulting and folding, junctions of continental and oceanic margins and along the plate boundaries. The main line of weakness however runs in three belts:

(1) The Circum Pacific Belt

It surrounds the Pacific Ocean like the circumference of the circle. Starting from the southern tip of south America it spreads along the Andes and Rocky mountains to Alaska. From Alaska it passes on to eastern Siberia through the Peninsula of Kamchatka and the Islands of Japan, Philippines, New Guinea, Newzealand come within this very belt, till it terminates in Antarctica. This belt of earthquake is identical with the chain of volcanic activity. This belt accounts for about 65 percent of the total earthquakes of the world. Japan alone records about 1500 seismic shocks every year.

(2) The Mid-Continental Belt

It is also known as Mediterranean belt or Alpine-Himalayan Belt. About 21 percent of the total seismic events of the world are recorded in this belt. It includes the epicenters of the Alpine Mountains and their offshoots in Europe, Mediterranean Sea, Northern Africa, Eastern Africa and the Himalayan Mountains and Burmese Hills.

(3) Mid-Atlantic Ridge Belt

It includes the epicenters located along the mid-Atlantic Ridge and several islands nearer the ridge. This belt records moderate and shallow focus earthquakes which are essentially caused due to the creation of transform faults and fractures because of splitting of plates and their movement in opposite directions. Thus, the spreading of sea floor and fissure type of volcanic eruption cause earthquakes of moderate intensity. Roughly the above mentioned seismic belts coincide with the volcanic belts of the world and more closely with the young mountain belt of the world.

EARTHQUAKE WAVES

These waves originate from the focus radiate in all directions in concentric circles and according to their mode of travelling and rate of movement they may be said to be of three types.

(1) Longitudinal Waves

These waves are also known as Push Waves Primary Waves or (P) Waves. These waves make the particles vibrate in the direction of their movement. They move through the earth but their rate of movement increases with greater density of medium through which they pass. Their rate of motion is 5 to 9 miles per second when they proceed through the central mass of the earth.

(2) Transverse Waves

These waves like the push waves move through the earth. But the only difference between the two waves is of speed as these are slow in movement than the push waves and cannot pass through liquid substance. These waves are also known as Secondary Waves, (S) Waves or Shake Waves and they proceed at right angles to the longitudinal waves.

(3) Surface Waves

These waves are known as surface waves long waves or (L) waves because they travel over the surface of the earth. These waves reach at the place of shock after making a full round of the earth. These waves are most terrible and cause greatest disaster. They cover the greatest distance as compared to the other two waves.

The magnitude of these waves and their place of occurrence is measured by an instrument called seismograph. The intensity of the waves and the distance from the place of origin can be calculated with the help of this instrument.

EFFECTS OF EARTHQUAKES

Apart from the immediate destruction of life and property, earthquakes produce some geological effects. Due to the passage of the surface waves fissures are formed through which water overfloods the region. It brings enormous quantity of sand and often produces sand dikes in these fissures. The roads are fissured, railway lines are twisted, submarine cables are damaged and bridges fall off from the pillars. A marked change is effected on the drainage system of the area. Sometimes the rivers change their courses. The new springs appear whereas the old ones stop. Sometimes the discharge of the springs increases and sometimes it decreases. Rock slides occur in hilly regions and bring down huge quantities of rock material. Sometimes the grounds are tilted and changes of ground level are at times effected. Relative displacement of areas and increase in the height of hills are also at times effected by earthquakes.

The earthquakes are sometimes no doubt highly disastrous to human life and property but after these harmful effects they are at least of some uses to man which are as under:

(a) The earth submerge the coastal land to form bays and inlets useful to man for navigation and fishing.

(b) New spring are formed which help irrigation etc.

(c) Big earthquakes have brought about upheavals of land to form hills and mountains which check rain bearing winds to cause rain. Sometimes new land for habitation also comes up.

(d) Lava accompanies with earthquakes which gives rise to the fertile soils.

(e) Sometimes, precious minerals and metals come up to the surface of the earth which are very useful for man.

VOLCANOES

According to P.G. Worcester (1948) a volcano is a vent, opening, shaft, channel or passage (circular in form), through which heated materials consisting of gasses, water, liquid lava and fragment of rocks are ejected from the highly heated interior to the surface of the earth. Prof. D.L. and A. Holmes (1978) are of the idea that a volcano is a vent or a line of communication between the earth's surface and the interior, from which flow of lava, foundation of incandescent spray or explosive bursts of gases and volcanic ashes are erupted at the surface. The mouth or the volcano is known as crater, which is a vast depression around it. The matter is accumulated in a conical hill shape. Examples are Fuji Yama in Japan, Vesuvius in Italy. The volcanic hills or mountains consisting mainly of lava tend to be dome-shaped rather than conical.

The material that comes out of a volcano is usually of three kinds-liquid, solid and gaseous. The liquid material is the most important and is known as the lava. This is nothing but molten rock and is composed of minerals. Accordingly as it consists of a larger or lesser proportion of silica, lava is said to be of two types:-

(a) With most silica or Acid lava.

(b) With less silica or Basic lava.

The acid lava is pale in colour, light in weight and usually melts at a relatively higher temperature. It is a thick fluid and moves slowly. On the other hand, the basic or basaltic lava is heavier in weight and darker in colour. A relativley lower temperature is required to melt it and being thinner in texture or more liquid it flows at a faster speed. The lava, as it comes out of the surface of the earth, contains some gas also, begins to cool and congeal. The basaltic lava solidifies into glass like sheets and the acid lava solidifies into a rough surface. Whenever a volcano erupt, vast quantities of dust, cinder or ash are also thrown up.

The volcanic dust are very fine particles, grey in colour and very light in weight. The solid fragments smaller than the finger joints thrown out by volcanoes are known as volcanic ash or cinder.

The liquid and solid material is usually accompanied with a large amount of gases. As soon as the eruption takes place a dark cloud of smoke and gases rises up in the atmosphere. These gases seldom produce any visible effect on the volcano and get dissipated in the atmosphere and effect the nearly rocks.

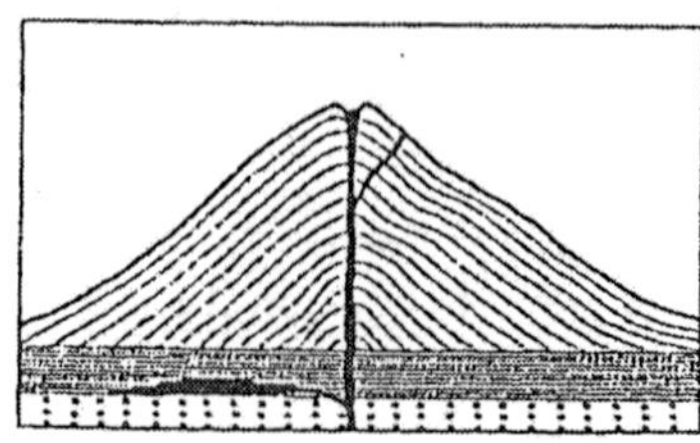

Fig. 12. Section through a Volcano, showing Volcanic neck, sill, and dyke.

DESCRIPTION OF VOLCANIC STRUCTURE

A typical volcano is generally conical in shape. The slope may be gentle at times and steep at other times. At the top of the cone there is a circular pit (Crater). Its diameter varies from a few hundred feet to several miles. The formation of both crater and cone is due to the accumulation of the products of explosive volcanic eruptions. The larger fragments of rocks fall near the pipe or mouth and form gradually a conical ridge round the pipe. As a result, the central region becomes depressed to form a crater. Thus cone and crater structure is very typical aspect of a volcano.

THE CAUSES OF VOLCANIC ACTIVETY

The exact causes of volcanic activety are not yet known. However, some of the probable causes are given as follows:

(a) Water

The water on the earth's surface seeps into the interior through many cracks and fissures. When it gets into contact with hot lava it turns into steam and then escapes out of the many holes and explosion is caused. Later on, lava also comes out of the holes.

(b) The Effect of Folds

When orogenetic action effects the earth's surface, the sedimentary rocks rise in folds. The pressure on lava is reduced at places which are the zones of weakness. At these places lava is melted and explodes as it forces itself out of fissures. This is why volcanoes are found close to fold mountain ranges.

(c) Radioactivity

A study of magma shows that radioactive particles are present in it. The heat generated by these radioactive particles melts the magma and forces it out of fissures.

TYPES OF VOLCANO

The Volcanoes during eruption throw out different types of material on the surface of the earth. It accumulates around the crater and acquires different shapes and sizes of cones, called accumulation mountains. Thus, on this basis we have the following three types of volcanoes and volcanic mountains.

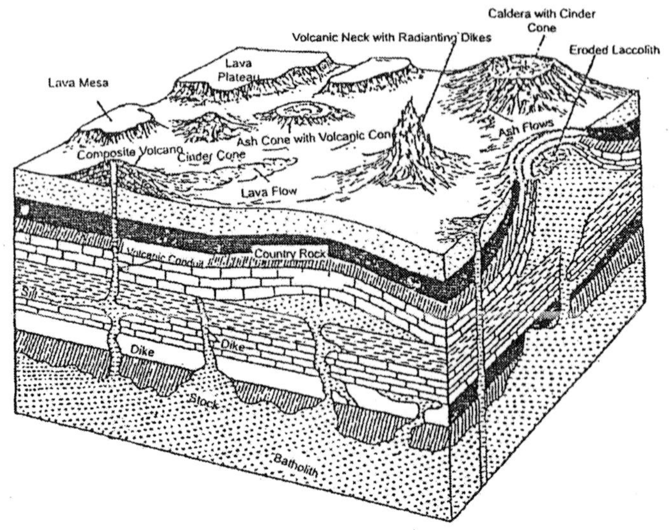

Fig. 13. Different types of landforms produced during volcanic activities.

(1) Cinder Cones

The conical structures are built around a volcanic pipe by the accumulation of the loose solid material, cinder and ash. The steepness of the cones varies with the nature of the material. The coarser material will form steeper cones while the finer material will produce low and gently sloping cones. The size of the crater is closely related to the frequency and violence of explosions. If they are short in duration but quite violent in intensity, a low broad cinder cone will be formed and vice versa.

(2) Lava Dome

This is a dome-like structure formed with an effusive flow of lava. But the shape of the dome varies according to the nature of the lava. If the lava is acid, the dome would be higher and less extensive with steep slopes. But if the lava is basic, the dome would be flattened over a larger surface, with gentle slopes, known as Hawaiian type or Shield cone. There are two sub-types of lava dome made by acid lava.

(a) The Strombolian Type

In this formation the crater is pipe-like and narrow.

(b) The Volcanian Type

In this structure the crater is much larger, having a funnel-like form.

Some times a larger part of the cone is destroyed and instead of the funnel shaped crater, a large shallow basin-like opening is formed, which has a flat floor, known as caldera. These are quite common in central France, Germany and Pacific Ocean region.

(3) Composite or Mixed Cones

When the effusive flow of lava alternates with explosive eruptions of cinder and ash, the composite cones are formed. During the explosive phase fragmental products accumulate round the pipe. During the quite phase of eruption lava piles are poured over these fragmental materials. Such a combination produces mixed types of cones. They are also called strata volcanoes for their rude or rough stratification.

PERIODICITY OF ERUPTION

The volcanoes of the world are divided into three types on the basis of period of eruption. They are said to be active when eruption occur frequently; dormant when no eruption has occurred over a long period of years; and extinct when there is no indication of future eruption.

(a) Active Volcanoes

The volcanoes which erupt frequently and constantly the molten lavas, gases, ashes and fragmental materials are known as active volcanoes. Barren Island in Andaman group of islands in our country and Etna and Stromboli of the Mediterranean Sea are the most significant examples. It is known as the light House because of its continuous emission of burning and luminous incandescent gases. Most of the active volcanoes are found along the mid-oceanic ridges. The latest eruption took place from Pinatubo volcano in June 1991 in Phillipines.

(b) Dormant Volcanoes

All volcanoes do not erupt regularly. Some of them may not show any sign of activity for years together nor any type of indication for future eruptions. But they suddenly erupt very violently and cause enormous damage to human life and property. Visuvious (Naples in Italy) volcano is the best example which erupted first in 1579 A.D. and then in 1631 A.D. The subsequent eruptions occurred in 1803, 1872, 1906, 1927, 1928 and 1929.

(c) Extinct Volcanoes

A volcano which is neither active nor dormant and has not erupted within the last few thousand years is called as extinct or dead volcano. This type of volcano has no indications of future eruption, thus the huge depressions on the top of volcanic mountains (craters) are filled with water and the lakes are formed. Mount Kilimanjaro in Tanzania (Africa), Lonar lake in Maharashtra and Kailash lake in Jammu and Kashmir (India) are the typical examples of extinct volcano.

VOLCANIC ERUPTIONS OR VOLCANISM

The effect of pressure increases with depth in the earth's interior. The pressure effect keeps the sub-crustal region in a viscous state. Any release of pressure which may be caused by the diastrophic movements will melt the rocks below and thereby pockets of magma will be produced in the sub-crustal region. This is one of the reasons for the c :currence of volcanic belts near the fold mountain belts of the world and the occurrence of igneous activity accompanying major crustal deformations. The magma when formed, will find its way upwards by the pressure of the overlaying rocks through fissures, joints, fault planes or bedding planes in the regions above or by melting and digesting the rocks above.

The origin of volcanism is due to the release of pressure at depth and it can be dealt with the explanation for:-

(i) The origin of magma with its high temperatures.

(ii) Origin of volcanic gases.

(iii) The extrusion of magma.

(i) Origin of Magma

It is believed that the central type of eruptions drains pockets of magma while the fissure type of eruptions may have some connection with the basaltic layer below the crust. Pockets of magma may be produced by the relaxation of pressure locally. Accumulation of radio-activity generated heat may also produce pockets of magma by the melting of rocks. This will also account for the heat. The heat may also come from the hot interior, as it is common experience that the temperature increases with depth. This heat is in a sense original, being the part left after so many million years of radiation.

(ii) Origin of Volcanic Gases

Steam, the most important of the volcanic gases, may result from the descending surface water through the earth's surface or the upper part of a volcanic cone. This stream may have magmatic origin as well. Some amount may be formed by the union of hydrogen of the magma with oxygen or more probably an oxide. The carbon dioxide may be produced by the heating effect of magma in a limestone region. Other gases like nitrogen, hydrogen, sulfur chlorine are formed within the rocks or come from descending waters.

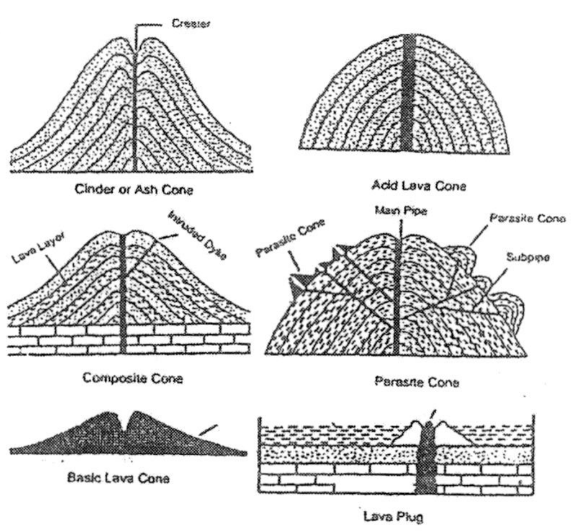

Fig. 14. Different types of volcanic cones.

(iii) Extrusion of Magma

Crustal deformation may produce magma on the one hand and on the other it may produce folding and faulting in the upper region and this will lead to the formation of numerous cracks, joints and fissures in the overlying region. These fissures serve as the paths of ascension of the magma. Magma is forced through these cracks by the weight of the overlying rocks. The imprisoned gases are also contributing factors to thus point. The energy of the gases will take the magma upon the surface just as water is forced up when a soda water bottle is uncorked.

VOLCANOES AND MAN

Volcanoes inspite of being quite destructive and harmful for human life and property are also useful to man in the following manner.

(a) The volcanoes help man in understanding the history of the interior of the earth.

(b) The lava, which a volcano erupts yields a very fertile soil for good crops.

(c) The volcanoes bring up precious minerals and metals from the interior of the earth which are very useful for man.

(d) The volcanoes are responsible for different changes in the relief features of the earth. They bring up new lands for habitation as well as for agricultural use.

DISTRIBUTION OF VOLCANOES

There are several thousand volcanoes in the world. According to Sapper the number of active volcanoes among them is nearly 430, of which 275 are in the northern hemisphere while 155 fall in the southern hemisphere. In these some are on land but the greater number is in the sea. They are closely associated with the seismic belts and weaker zones (folded mountains) of the world. Like earthquakes, volcanoes are found in a well defined three belts or zones in the world viz. circum-Pacific belt, mid continental belt and mid-oceanic ridge belt.

(1) Circum Pacific Belt

This volcanic belt covers the coastal margins of Pacific Ocean.

It begins from Erubus Mountain of Antarctica and runs northward through Andes and Rockies mountains of south and north Americas to reach Alaska. Form Alaska it turns towards eastern Asiatic coast running through Japan coast running through Japan and Phillipines upto East Indies. Most of the high volcanic cones and mountains are found in this belt. Examples of famous volcanoes are Aleutian, Hawaii, Cotopaxi, Fuji-Yama (Japan), Shasta, Rainier and Hood (N.America), a valley of ten thousand smokes (Alaska), Mt. St. Helens (Washington U.S.A.), Mt. Tool, Pinatubo and Mayan of Philippines. On account of its continuous fire girdle this belt is sometimes termed as Pacific ring of fire or volcanic zones of the convergent oceanic plate margins.

(2) Mid Continental Belt

This belt with a east-west direction, starts from the east Indies where it cuts the former belt, passes through Asia Minor, the Mediterranean continues up to the West Indies. It includes the volcanoes of Alpine Mountain chain and Mediterranean Sea and the volcanoes of fault zone of eastern Africa. The famous volcanoes of Mediterranean Sea such as Stromboli, Visuvious, Etna and volcanoes of Aegean Sea are included in this belt. As contrary to the first this belt does not have the continuity of volcanic eruption as several volcanic free zones are found along the Alps and Himalayas. This is because of compact and thick crust formed due to intense folding activity. The important volcanoes of the fault zone of En. Africa are Kilimanjaro, Meru, Elgon, Birunga, Rungwe etc. This belt is also known as the volcanic zone of convergent continental plate margins.

(3) Mid-Atlantic Belt

This belt includes the volcanoes mainly along the mid Atlantic ridge. The most active volcanic area is Iceland which is located on the mid Atlantic ridge. Recently, Hekla and Helgafell volcanoes erupted in 1974 and 1973 respectively. Other active volcanic areas are Lesser Antillers, Azores, St. Helena. Dreadful and disastrous eruption of Mount Pelee occurred on 8th of May, 1902 on the islands of West Indies in Carbibbean sea. All the 28000 persons, except two persons were killed by killer volcanic eruption.

5
Mountain Building

The earth's surface is undergoing constant change. Such changes take place with dramatic suddenness, in case of earthquakes and volcanicity. But usually they are so gradual as to be almost imperceptible. The mountains of today may become the plains of tomorrow by erosional process and denudation.

There are two main forces which are responsible for these changes viz.

1. Endogenic
2. Exogenic.

(1) Endogenic forces are further sub-divided into —

  (a) Slow forces, which produce tension, compression and result in the formation of folds and faults.

  (b) Sudden forces, which produce pressure in magma in the interior of the earth and cause earthquakes and volcanic eruption.

(2) Exogenic forces are connected with atmosphere or agents of erosion such as rain, snow, wind and sun. The exogenic forces are destructional or erosional forces while the endogenic forces are the constructional forces which tend to form the land masses. Hence, These great mountain building forces or movements result in the formation of mighty ranges and mountain chains. Being subject to both these forces the land rises and sinks, islands appear and disappear and the face of the earth being continually changed.

MOUNTAINS

A mass of land projecting well above its immediate surroundings is called a mountain. This is the most prominent feature and uplifted portion of the earth's surface. It has sloping sides and a summit or peaks at the top. There is no fixed boundary at which a hill is regarded as mountain. One criterion is that any feature higher than 6600 feet above sea level is a mountain but generally a feature is regarded as a mountain if it extends at least 3300 feet above the sea level.

Being of complex character, the mountains of the world have some common characteristics which are as under:

(a) All the mountain system consists of stratified rocks which show that these must have been formed in the bed of some sea.

(b) It has been found that the thickness of the strata in the folds or mountains region is more than plains. Moreover, the thicker folded stratas are greater in length but very much narrow in their width.

(c) The mountain systems are usually arc-shaped and run east-west or north-south.

According to Suess, the arc shape of the mountainous systems is the outcome of outward push of the rock strata. This push started from the centre towards the outer rims, giving it a wave like structure and are likely shaped.

TYPES OF MOUNTAINS

Mountains may be generally classified according to their mode of formation into following three types.

(i) Accumulation type, formed by the volcanic eruptions such as (a) cinder cone; (b) lava dome; and (c) composite cones. Volcanic mountains and sand dunes are the examples of accumulation mountain which has already been discussed in this chapter under the heading 'Volcanoes'.

(ii) Relict or Residual type, formed by differential erosion. This is specially noticeable in regions which are composed of materials of differing solubility. As soon as a mountain comes up, the streams, glaciers and winds start to reduce its size and extent. The plateau remnants (Chotanagpur plateau) and monadnocks are the examples of relict type of mountains.

(iii) Deformation type, formed by crustal movements such as (a) folded mountains; (b) block mountains; and (c) dome mountains.

(a) Folded Mountains

In the past geological ages and even now the earth's crust is in a most unstable condition. When the strata is compressed from side to side the beds are bent into great up folds (anticlines) mountain and downfolds (sysclines) longitudinal valleys. Because of these troughs and arches, the whole crust appears like a wave. Generally the contraction or push is quite intense and from all sides, which results in various shapes and sizes of folds. The accompanying figures will illustrate the different forms of folds.

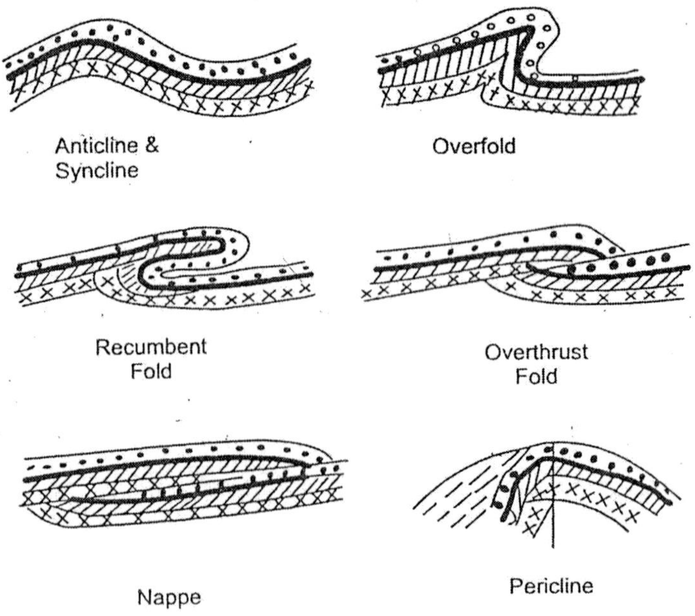

Fig. 1. Various forms of folding.

TYPES OF FOLDS

(1) Symmetrical Fold

This is the simplest type of fold open and upright in which an anticline or a syncline is symmetrical about the axial plane which is vertical in it. It is also called the Jura type.

(2) Asymmetrical or Inclined Fold

Such folds are inclined more or less in one direction along the axial plane and the symmetry is lost in this type of mountain.

(3) Overturned or Overfold

In this type of folding one limb or side of an anticline or a syncline is inclined to a great extent towards the other.

(4) Recumbent Fold

This type of folding shows one limb practically resting on another and the two limbs are more or less in a horizontal position. In this type the plane of the axis forms a flat angle and one side of the fold lies parallel to other due to strong compression.

(5) Isoclinal Fold

This type of fold shows parallel limbs which may either be vertical or inclined.

(6) Monocline Fold

In this type of folding the strata are bent in one direction only. On both sides of the bend strata are horizontal.

(7) Homocline Fold

If there are several beds all dipping in one direction, the resulting structure is called a homocline.

(8) Fan Folding

Fan Fold refers to an 'anticlinorium', having the axial planes of folds converging towards the centre. Such folds resembles a fan, consisting of several minor anticlines and synclines.

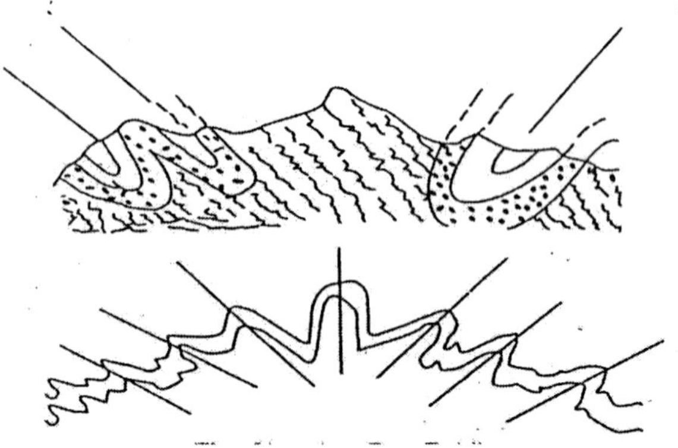

Fig. 2. Showing Fan Folding.

(9) Open Folds

Open folds are those in which the angle between the two limbs of the fold is more than 90 degree. Such folds are formed due to moderate nature of compressive force and give a wave-like scene.

(10) Closed Folds

Closed folds are those folds in which the angle between the two limbs of a fold is acute angel. Such folds are formed because of intense compressive force.

(11) Nappe Folds

Nappe refers to a huge recumbent fold having both limbs practically horizontal that has been forced to a great distance, i.e. several kilometers, over the underlying formation thereby covering them like a cloth. The word 'Nappe' has been derived from the French, meaning the table cloth.

(b) Block Mountains

In addition to forming upfolds the crust of the earth also cracks under the strain of compression or push. Often where such cracks occur the portions on either side slip against each other; one part rising up and the other slipping down. This slipping movement is known as faulting. Elevated mountain masses formed in this way are known as "block mountains" or horst. The central plateau of France and the Meseta of Spain are best examples of this type.

In other cases the strata between two parallel faults subside to form a rift valley or graven, which thus lies between two parallel sets of block mountains. In this way was formed the Rhine Rift Valley of Germany between the Vosges and the Black Forest. The greatest rift valley in the world is that stretching from the Jordan-Dead sea trough, through the Red Sea to Lake Nyasa, in East Africa.

(c) Dome Mountains

It is a big upwarped structure corresponding to the crest of a wave form or an anticline which does not extend over a long distance. Domes may be produced by localised earth movements. These surface slopes or the beds dip from the central part in all directions. Dome-like, structures also result from spheroidal weathering as in the case of Archean rocks of India. Dome-like structures may also result from the accumulation of lava materials from volcanoes or by the intrusion of salt material in plastic state near the surface of the earth. Just opposite to the dome structure is the basin, a trough like formation resembling a syncline which does not extend over a long distance.

CAUSES OF MOUNTAIN BUILDING

Various explanations and ideas have been put forward to account for the uplift of the mountain masses. But compressional forces are mostly responsible for the formation of the fold type which is the most common type of mountain.

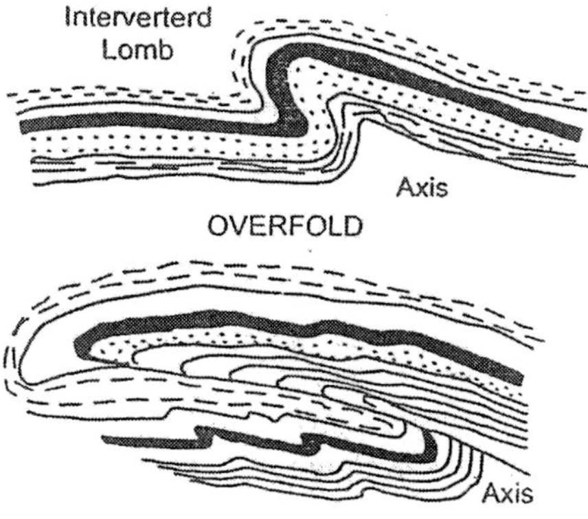

Fig. 3. Recumbent Fold of NAPPE.

Contraction Hypothesis

(1) This states that the earth through radiation of heat, is cooling and as a result the earth is shrinking and wrinkles developing on its surface. But with the discovery of the fact that the radio active substances produce heat on disintegration, this idea of secular cooling has lost much of its popularity.

(2) Some hold that under the load of sediments the floor of the subsiding geosynclinal basin, is likely to be broken and thus sediments would meet the internal heat and would expand in volume, as a result the upper sedimentary layers would be uplifted. This process, if in operation, would account for the vertical uplift only but not the compressional forces. This may come into operation during the late stage of mountain building.

(3) When the geoynclinal basin, on receiving the sedimentary load, will sink down, the two sides of the shallow trough will be brought nearer. This will generate compressional forces and may account for the formation of fold mountains.

(4) Isastatic adjustment is believed to play an important part in mountain-building but the process only accounts for vertical uplift and not the compressional forces. Its part, though considerable, may be held to be subordinate.

(5) Continental drift hypothesis accounts well for the mountain building process. The equator-ward force according to Wegner is responsible for the origin of the Alpine Himalayan Chain of mountains and west-ward force is responsible for the formation of Rockies and the Andes.

(6) Thermal Cycle Hypothesis: The cycle which is controlled by the radio-actively generated heat is called Joly's thermal cycle after the name of propounder. He states that the sail which constituted the continents is floating on sima a denser element. During the melting of sima, the sial layer will be in a state of tension and with the escape of heat, shrinkage of the sima, due to

solidification, will start and the sial layer in order to accommodate itself to the shrinking siesmic layer will be thrown into folds.

(7) Convection Current Hypothesis: According to Holmes there is a convection current from the equator to the poles in the sub-crustal region which causes crustal deformation is responsible for the mountain building. The radio-actively generated heat in the seismic layer will make it molten and will allow the flow of convection currents.[30]

IGNEOUS ACTIVITY DURING MOUNTAIN BUILDING

Formation of complex fold mountains is generally associated with the intrusion and extrusion of magma. The magmatic intrusion and extrusion may be due to the increase and reduction of pressure as a consequence of the uplift of the sedimentary pile. Batholiths of granite are generally associated with complex fold mountains. These granitic cores are sometimes laid bare by deep erosion. The batholiths are generally associated with sills, dykes etc. the formation of which is due to the intrusion of magma into the planes of weakness in the mountain rocks. Sometimes even intrusion of magma takes place with the resultant volcanic activity.[31]

IMPORTANCE OF MOUNTAINS

Mountains are the source of considerable wealth such as the source of timber, fuel, woods, herbs etc. Several types of minerals (gold, silver, copper, tin) are mined from the mountains. The great mountain system of the world are mainly important for their minerals and in the temperate zone for their timber. In the plateau regions of some mountain systems agriculture has been made possible by irrigation and above the forests in temperate areas there are valuable alpine pastures. The swift streams of mountains are frequent sources of hydroelectric power, especially in countries which have no coal, such as Switzerland and Norway.

In north America, the western Cordillera provides gold, copper, lead and silver, especially in the states of Nevada and Montana. The Andes provide tin and copper (Bolivia) gold and platinum (Colombia) and silver (Peru). The Highlands of East Australia are important for copper and gold.

The lumbering industry is specially important in British Columbia, Washington and Oregon (soft woods), the central American mountainous lands (hard woods), the Himalayan slopes (teak and sal) and Scandinavian mountains (soft woods).

Mountain pastures have been utilised most extensively for cattle rearing in Switzerland and Scandinavia. The mountains are effective climatic barriers and the climates of regions on either side of a high mountain range are very different. Such as coastland of British Colimbia have an equable climate and heavy rainfall. While the lands to the east of the Rockies have an extreme climates. Again, the climate of the mountainous areas differs from that of the adjacent lowlands.

High mountain ranges are also barriers to communication and so tend to separate peoples. Traffic across mountains is limited to the passes, which are often so high as to be snowbound in winter. Such ranges as the Alps, Andes etc. can only be crossed with great difficulty or by expensive tunneling. Mountains are great source of waters as snow melts of high peaks and give rise to rivers. Mountains have a scenic beauty and health resorts because of their cool climate, thus attracts tourists and religious people. [32,33]

ROCKS

The top or the crust of the earth on which continents and oceans rest, is called the lithosphere. The word lithosphere, infact means rocksphere as the literal meaning of 'Lithos' is rock. Geologically speaking, all materials that make up the crust of the earth are rocks, whether they are hard like granite boulders, combustible coal, soft like clay, loose fragments of gravel and sand, compact like slate and porous like limestone and chalk.

The smallest component of the crust of the earth is element. The eight abundant elements viz. iron, oxygen, silicon, magnesium, nickel, sulphur, calcium and ammonium constitute 99% of the total mass of the earth, in which first 4 account 90%. One element is organized to form compounds known as mineral. The important mineral groups are silicates, carbonates, sulphides, metal oxides etc.

The outstanding rock forming silicate minerals are quartz, feldspar and ferromagnesium. Feldspar is used in ceramics and glass industry. Carbonate group of minerals is susceptible to chemical weathering and erosion in humid regions. Limestone and marble having calcite are corroded by ground water and extensive caves are formed. Sulphide minerals include pyrites, iron sulphides when comes in contact with water or air forms sulfuric acid and causes serious environmental problems. Metallic elements like iron, aluminum etc. after reacting with atmospheric oxygen form metal oxides, which are commercially very important.

Rocks play very important role in determining the characteristic features of landforms as the magnitude of erosion largely depends upon the structure and composition of rocks. Rocks are naturally occurring aggregate of minerals and are units with which the earth's crust is composed. They are pages upon which the earth's history is written and a study of them is essential for the proper understanding of the science of the earth.

TYPES OF ROCKS

There are various basis on which the rocks can be classified but the most fundamental and comprehensive is the classification of rocks on the basis of their origin. The rocks are classified into three broad categories on the basis of their mode of formation viz. (1) igneous, those formed by the agency of fire; (2) sedimentary, formed by water; and (3) metamorphic (changed form of first two), formed by heat, water and pressure at depths.

(1) Igneous Rocks

The earth started its career in a state of burning gases. In its revolution through space it began to radiate heat and consequently began to cool. From the gaseous state the earth first passed into the liquid state and then into the solid state (at least on the upper surface). This solidified rock type is the igenous variety. The word igneous is derived from the Latin word ignes meaning fire. This is because of the fact that such rocks are formed from molten and heated material. The molten rock material together with the gas content, is called the magma. Igneous rocks are formed from the solidification of magma.

CHARACTERS OF IGENOUS ROCKS

These rocks have been formed from the solidification of very hot molten rock material called magma. Such rocks generally originate at depths but sometimes are formed on crust of the earth. Others are formed when their magmas are arrested in their upward ascent somewhere in the crust and solidify there. Upon this there have been recognised two distinct classes of igneous rocks:

(a) Intrusive and

(b) Extrusive

(a) Intrusive Rocks

When an igneous rock has cooled quite slowly at a very great depth inside the earth, it is known an intrusive rock. Granite, entirely crystalline in nature is a typical example of this type.

The intrusive rocks are further sub-divided into (i) Plutonic and (ii) Dyke rocks. This classification has been given on the basis of place of cooling and time taken in cooling by the molten matter.

(i) Plutonic Rocks

Plutonic rocks are those which have solidified deep down below the surface of the earth. They are highly crystalline in nature as the gases cannot escape and the cooling is very slow. They are seen only after long continued deep erosion.

(ii) Dyke Rocks

Some of the molten material which leaves the interior of the earth but due to certain reasons cannot reach the surface of the earth. This material enters the dyks and fishures of the rocks and cools down there, giving origin to dyke rocks, which is always crystalline in nature.

(b) Extrusive Rocks

When an active volcano erupts the lava on the earth's surface it cools down rapidly and no crystals are formed. This type of cindery and pumice stone rock which comes into existence is known as extrusive. Basalt is the typical example of this type of rock.

In volcanic type of rocks, there is no effect of pressure and the gases also escape easily. So the material solidifies at a faster rate forming the coarse grained rock with phaneritic texture. These rocks do not contain any fossil nor do they show any clear stratification.

COMMON IGNEOUS ROCKS

Granite

It has acquired its name from its well developed grains, being most common variety of igneous rock. It is medium to coarse grained and generally light coloured. It consists of quartz and feldspar. Micas, both muscovite and biotite are frequent.

Pegmatite

This rock occurs at dykes and veins of the country rocks. During the late stage of crystallisation of the magma, the residual part becomes generally very rich in volatile matter, alkali and silica. Commonly quartz, orthoclase and mica are present besides topaz, apatite and lepidolite to some extent.

Syenite

The name has been derived from Syene in Egypt. It is also a medium to course grained rock, very similar to granite. The distinction between the two lies in their mineral and chemical composition, as the silica percentage being lower than granite.

Diorite

It is a medium to coarse grained rock generally darker in colour than preceding ones, having lower alkali percentage.

Gabbro

It is a plutonic rock of dark colour and low silica percentage. It has plagioclases and pyroxenes minerals contents are frequently in it.

Dolerite

It is a medium to fine grained rock, having chemically the same composition as of gabbro.

Basalt

It is the most common volcanic rock, with a fine grained texture and dark colour. Chemically it contains lime-rich plagioclases and pyroxenes as the primary minerals. The other contents being iron ores, sphere, ilmenite etc. The word Trap derived from Swedish word trapf meaning stair, is a general term applied to basalts because they often present a stair-like aspects on weathering. Deccan Trap is a typical example.

Peridotite

A course grained dark coloured rock, having mostly the ferro-magnesian minerals.

(2) SEDIMENTARY ROCKS

When the surface of the earth was sufficiently cooled to hold water, oceans and other areas full of water came into existence. From that time began the destruction of the up-standing masses by the different agencies of erosion. The products of erosion of the denuding agents were brought to lakes, seas and oceans and by the consolidation of such sediments, sedimentary rocks were formed. This is about the first formed sedimentary rocks. Later sedimentary rocks have been formed from sediments derived from other igneous, sedimentary or metamorphic rocks.

The sediments eroded by rivers, glaciers and wind are re-deposited in the beds of oceans and seas in the form of layers, so also known as stratified rocks or layer rocks. These layers are compressed under the weight and cemented by lime and other fine material become hard and compact. Sandstone, limestone, coal and slate belong to this class of rock.

CHARACTERS OF SEDIMENTARY ROCKS

The sediments of these rocks are the products of erosion both mechanical and chemical from some pre-existing rock masses. These sediments are carried by the running water both in suspension as well as in solution and are deposited in basins or sea beds. The marine deposits also contribute the accumulation. The conspicuous layering is the chief characteristic of the sedimentary rock, which may be either horizontal or inclined. The sedimentary rocks contain fossils (remains of animals and plants) in it. Each individual layer is called stratum and separating plane between two layers is known as bedding plane.

COMMON SEDIMENTARY ROCKS

Sandstone

This is the rock formed by cementation of sand grains to hard mass. Silica generally constitute the main mass. It is of different colours depending on the cementing material. They vary in size such as boulder, cobble, pebble, gravel, sand, silt and clay.

Limestone

This is a sedimentary deposite of calcium carbonate. The deposition is either from solution or from remains of the dead bodies of marine animals. Corals are the chief contributer of calcium.

Marl

It is an impure variety of limestone, with clayey material. It is sometimes used as cementing material.

Dolomite

It contains a good amount of magnesium carbonate. Limestone often change to dolomites.

Conglomerate

It is a sedimentary rock consisting of rounded and sub-rounded fragments, usually water work pebbles, cemented together by a matrix of calcium carbonates, silica etc. It is a consolidated gravel.[35]

Chalk

It is a soft and loose variety of limestone in which sometimes foraminifered shells are aboundant.

Coquina

It is a variety of shelly limestone.

Shale

It is formed by the consolidation of clay and mud.

Argillite

This is a variety of clayey rocks which is harder than mudstone, in which shelly structure is absent.

Fireclay

It is a variety of clayey rock which is commonly found in coal fields underlying the coal seams. It can withstand high temperatures without melting. It is used as refractory substances.

Siltstone

It is a variety of sedimentary rocks which is formed by the consolidation of silt materials.

(3) METAMORPHIC ROCKS

The igneous and sedimentary rocks later on with the passage of time may be subjected to heat and pressure accompanying crustal movements. The effect of these is to change the character of the pre-existing rocks and in this way the metamorphic rocks result. Metamorphosis is a Greek word meaning- change, so the altered forms of igneous and sedimentary rocks are the metamorphic rocks. Coming under the influence these forces, particularly during intense earth movements, their original character and appearance may be greatly altered. Thus the sedimentary rocks often become crystalline in appearance and are sometimes difficult to distinguish from igenous rocks. By the processes of metamorphism chalk and lime stones are changed to marble; clay to slate; sandstone to quartzite; shale to schist; granite to gneiss and coal to graphite and anthracite.[34]

CHARACTERS OF METAMORPHIC ROCKS

The metamorphosed means changed, generally result from the changes brought about by heat, pressure and chemical agencies, upon pre-existing rocks. Thus, those derived from igneous rocks are called arthometamorphic and those derived from sedimentary rocks are parametamorphic rocks. There is some parallel arrangement of minerals (foliation) which distinguishes them from igneous rocks with a peculiar texture. These minerals are kyanite, sillimanite, zoisite, wollastonite, staurolite and andalusite Due to metamorphism igneous and sedimentary rocks are totally changed in their physical state, chemical composition and crystallization of minerals. According to the forces that change the rock metamorphism may be said to be of the following two varieties.

(a) Contact Metamorphism

It is also known as local metamorphism. It is a process of change effected by heat in the proximity of magmatic intrusion in the effected rocks. The heat is produced by both the heat of magma and chemical constituents. This metamorphism produces chemical replacement, re-crystallization and mechanical changes in the host rock. As the host rock gets metallic substance from the magma so it becomes a great source of valuable minerals. Marble, quartzite, slate and graphite are the examples of such metamorphism.

(b) Regional Metamorphism

It is also known as dynamic or tectonic metamorphism. In this process the original rocks are transformed mainly under pressure but partly under temperature at a great depth of the earth's crust. The deformation of crystal increases the pressure in the effected region. It is associated with mountain chain. Slaty and plastic deformation are the two important results besides re-crystallization in regional metamorphism. Gneiss and schist are example.

COMMON METAMORPHIC ROCKS

Gneiss

It is a common coarse grained rock which shows a banded appearance. Quartz, feldspar and mica are the generally occurring minerals. Sometimes granite changes into gneiss.

Marble

It is the changed form of limestone. Marble may be of many colours, but generally it is white or light coloured. It can be easily worked because of its softness and hence form very good ornamental building stones.

Quartzite

This is the metamorphosed equivalent of sandstone and consists mostly of quartz. They often occur as veins cutting across other metamorphic rocks, being usually very hard.

Slate

More pressure on clayey mud produces successively slate, phyllite and finally schist. The minerals are very fine grained in slates. The easy plane of fissility renders it very suitable for being used as roofing material in thin slabs.

Phyllite

These are produced with more pressure. They are more or less slate like in appearance. The only difference is that phyllites have a glossy and shinning appearance which is due to the formation of incipient mica flakes.

Schist

Schist is formed from the metamorphism of shale. Mica, quartz, chlorite, kyanite, graphite and hornblende are the constituent minerals of shcist.

Laterite

The word has come from the Latin word "later", meaning brick. This is brownish in colour and is mainly composed of iron, aluminum and oxide of manganese.

Black Cotton Soil or Regur

It results from the decomposition of basalts, containing a high percentage of the oxide of calcium, magnesium iron and alkali, hence it is very fertile. It occurs in Maharashtra and adjacent areas and is extensively cultivated for cotton growing.

WEATHERING OF ROCKS

Ever since the consolidation of the earth's crust, the rocks have been subjected to the wearing and tearing effect of weathering agents. Weathering involves the process of breaking of exposed rocks to smaller particles and decomposition of rocks, caused by changes in temperature, frost action, plants, animals and man. The weathering does not change or disturb the place of rock material. It involves three distinct processes as follows.

(1) Physical Weathering

This is also known as mechanical weathering as it refers to the mechanical disintegration of rocks in which their mineral composition is not changed. It is done by solar heat, rainfall, frost, wind, vegetation, man and animals.

(2) Chemical Weathering

It refers to chemical decomposition and disintegration of rocks. It includes oxidation, hydration, carbonation and solution. The agents of process are water, humidity and the various gases present in the air. It is more active in tropical regions than in temperate regions.

(3) Biological Weathering

A process of weathering in which the rocks are destroyed by plants, animals and man. The long and tenacious root fibers of plants go down into the cracks of rocks and dislodge them. The earthworms, ants and rats etc. make borrow through the rocks and destroy the main composition of rocks. The excretion of many of them provide acids for a gradual decay of the rock. The quarrying, mining, deforestation and indiscriminate cultivation of land by man also weathers away the rocks.

CONTROLLING FACTORS

The nature and magnitude of weathering differs from place to place and rock to rock. Weathering of rocks is effected and controlled by its agents, rock structure, height, slope, climatic conditions and topography to a greater extent. As for example the disintegration of rocks is more effective in hot and dry regions and frost regions, while chemical decomposition is more prevalent in humid and temperate regions.

(a) Composition of Rocks

Carbonate rocks e.g. calcium, magnesium, having more soluble minerals are easily effected by chemical weathering. Rocks having vertical strata are easily loosened and broken down due to-temperature changes, frost action, water and wind.

(b) Nature of Slope

The rocks of steep hill-slopes are easily disintegrated due to mechanical weathering. Natural removal of weathering material along the hillslope (rockfall) allows continuous exposure of rocks to atmospheric conditions for further weathering.

(c) Climate Variations

Because of abundance of moisture and high temperatures, leaching process and solution of rocks (salt, limestone etc.) are more effective in the humid tropics. In semi-arid regions the rocks are weakened due to alternate expansion on heating during daytime and contraction on relative cooling during nights because of diurnal change of temperature. Limestones, very weak rocks in humid regions, become more resistant to weathering in hot desert climate.

(d) Effect of Vegetation Cover

Infact, vegetations, bind the loose rocks with their roots and thus protect them from weathering. Dense vegetation protect the ground surface from direct impact of sun rays. On the other hand, the penetration of thick roots of trees weakens the rocks by breaking them into several blocks.

6
Denudation, Weathering and Erosion

DENUDATION

The process of levelling down or wearing away the land forms of the earth is called denudation. Denudation which also includes the removal of the materials worn away by the elements of weather, gradually lowers the elevation of the land. In other words, as soon as, a new landform emerge the denudation process starts. It includes two processes, firstly the "weathering" or breaking up of rock material and secondly the "erosion" or carrying away of the material to lay the rock bare for further attack. The erosion is accomplished by the agents of running water, moving ice, wind and gravity, whereas the rain, frost and temperature are chief agents of weathering. Both rocks and soils wear away at differential rates because of differences in their internal composition and because of the variable nature of the environment around the world.[37]

To make it more elaborate, denudation is the wearing away of the land by various natural agencies viz. the sun, wind, rain, frost, river, glacier and the sea. The heat causes the rocks to expand, crack and break up; the wind carries loose particles to wear away the rock; the frost widens the cracks; the sea wears away the coasts; the rivers, glaciers and wind wears away the land and transport the eroded material to be deposited elsewhere. Denudation is thus one of the two major processes responsible for earth sculpture or the change in form of the earth's surface, the other being deposition.[38]

WEATHERING

According to B.G. Sparks, "weathering may be defined as the mechanical fracturing or chemical decomposition of rocks by natural agents at the surface of the earth".

Arthur Holmes 1952 has presented more elaborate definition of weathering which also includes the processes of weathering. According to him, "weathering is the total effect of all the various subaerial processes that cooperate in bringing about the decay and disintegration of rocks, provided that no large scale transport of the loosened products is involved. The work of rainwash and wind, which is essentially erosional, is thus excluded."

The rocks go on decaying and breaking at their own positions, under the influence of climatic forces like temperature changes and precipitation. Weathering is, thus, directly or indirectly related to the atmospheric conditions. Insolation, frost, rain water, gases and organisms are the principal agents of weathering. These agents involve the process of weathering in two ways viz. mechanically (breaking and disintegration) and chemically (decomposition).

TYPES OF WEATHERING

There are three distinct types of weathering on the basis of its dominant agents, which would be studied one by one separately.

(1) Physical Weathering

It is also known as mechanical weathering which leads to fragmentation and breakdown of rocks into boulders and blocks, cobbles and pebbles, sands and silts and chemical decomposition of feldspar and mica to form clay. The chief agents of physical weathering are temperature, frost, wind and moisture.

(a) Temperature

In desert regions the diurnal range of temperature is generally very large, so the rocks are highly heated by the scorching rays of the sun, during the day time and consequently expands. During the night time reverse is the process, hence the rocks contract considerably. The outer layer, therefore, gets detached from the main mass and disintegrates due to the alternate expansion and contraction. Then again all the minerals do not expand to the same extent by heating. The differential expansion and contraction of minerals break them into different sizes of grains. The breaking of rocks due to insolation is in two ways block disintegration and granular disintegrated.

When the rocks are disintegrated along the joints and broken big blocks of rocks are dislodged from the main mass, it is called as block disintegration. On the other hand some coarse grained rocks are broken into smaller grains due to differential insolation caused by different coloures of rocks, are known as granular disintegration.

(b) Frost

Frost action alongwith temperature variation, is a major agent of physical or mechanical weathering. When water, which percolates and collects in the joints and cracks of rocks, freezes, it expands in volume by 9 percent. This increase in volume pushes out against the sides of the rock joint. This outward tremendous pressure may be sufficient to gradually break the rocks apart. More likely, the rocks are broken apart by a combination of freezing (thawing and stretching) and shrinking processes over a long period of time. Frost action is very conspicuous in regions of higher latitudes and over altitudes. The presence of joints and cracks helps the frost action to a considerable extent by allowing percolation of water inside the rock.

(c) Wind

Wind performs several actions which are mainly mechanical in nature and are clearly seen in arid deserts, semi-arid regions, and on bare coastal regions, where vegetation is sparse. The blown particles strike against upstanding masses and cause erosion by the mechanical wearing of the rocks, known as 'wind abrasion'.

The another action of wind is 'exfoliation', a process, where by thin eggshell - like rock pieces are detached from the main mass layer by layer. The breaking away of one layer exposes a new spheroidal surface for further wearing. The continual stretching and shrinking, as well as the slow chemical breakdown of minerals in rocks is the result of exfoliation. Igneous rocks such as granites and basalt are specially susceptible to this weathering, whereby rounded mountains called exfoliation domes, are formed.

(d) Moisture

The outer shells of the rocks are shattered due to sudden light showers mainly hot desert areas. The highly heated rocks when struck by sudden drizzles develop numerous cracks. The repetition of this mechanism causes spelling and granular disintegration of rocks.

(2) Chemical Weathering

It is a process wherein the minerals of the rocks weather away due to chemical reactions by disintegration and decomposition. The rain water and atmospheric gases are the media which activate several types of chemical reactions within the rocks. This type of decomposition and decay of rocks is common in hot wet regions of the world. The principal agents of chemical weathering are solution, oxidation, carbonation and hydration.

(a) Solution

Water has got both disintegrating and decomposing power as it is a good solvent. It can dissolve many mineral substances. Leaching is the term which is used for the removal of soluble substances in solution by percolating water. The chemical action of rain is due to the carbon dioxide which absorbs during its passage through the air and becomes acidic. Limestone and other forms of calcium carbonate, such as chalk and marble, are soluble in carbonic acid, besides gypsum and salt-rocks.

(b) Oxidation

It is a process whereby oxygen, dissolved in water after reacting with the rocks produces several types of oxides. When water mixed with atmospheric oxygen comes in contact with iron bearing rocks it shows a conspicuous colour changes by oxidation. The rusting of rocks weakens them and ultimately the rocks are disintegrated. The ferris oxides and ferris hydroxides give red and yellow colours to many rocks and soils as you see in rusting of iron.

(c) Carbonation

The rain water absorbs carbon dioxide from the atmosphere and also from decaying organic matter, becomes a good solvent of limestone and a producer of clay from feldspar. Limestone is especially susceptible to carbonation because the mineral calcite, readily dissolves into calcium carbonate, or lime and is rapidly dissolved from limestone.

(d) Hydration

Chemical combination with water is called hydration. It is often accompanied by slight increase in volume. This exerts additional pressure and acts like wedge similar to a certain extent to the frost action. Igneous rocks are very susceptible to this type of breakdown. The mineral feldspar, common in granite, forms into a clay mineral when combined with water. Quartz, a fairly resisting mineral is soluble in alkaline water. Hydration is an exothermic reaction and involves a considerable volume change which may be important in physical weathering - exfoliation and granular disintegration. Hydration prepares mineral surfaces for further alteration by oxidation and carbonation.

(3) Biological Weathering

Both plants and animals are largely responsible for the break down of the rocks. At present man itself has become the most powerful agent of weathering because of the modern development. Though vegetations protect the rocks by binding them through their roots but different types of acids e.g. humic acids, bacterial acids and microfloral acids, produced by them facilitate biochemical weathering. The main agents of biological weathering are plants, animals and man.

(a) Plants

Tree roots grow into rock joints and widens the cracks, thus promoting the disintegration of rocks. Plants also produce an organic product called humus after their decay which is very helpful in chemical decomposition of rocks.

(b) Animals

Burrowing animals can also trigger mechanical weathering. They can dig beneath a rock mass, weaken the rock's foundation and cause the rock to collapse under its own weight. Decaying organisms (bacteria) release acids that can assist chemical weathering.[36]

(c) Man

Man accelerates the weathering of rocks by deforestation, mining and blasting of hills for construction of roads and dams. He also makes the grounds for other weathering agents for further disintegration of rocks.

EROSION

When the fragments of rocks are dragged along the surface of the earth, it produces cuts, scratches, grooves and holes on it which is known as erosion. Thus, erosion is that process which removes the products of weathering agents. It is a broad term as and when compared with weathering. The breakdown of rock and soil in place caused by changes in temperature, frost action, plants and animals, is called weathering. It is a static process. On the other hand, erosion is a mobile process as it involves both the processes of gradation and aggravation, side by side. It does the work of plucking, cutting and chiselling; scraping, scratching and grooving; dragging and aggrading. Thus the main distinction between weathering and erosion lies in the fact that in erosion there is break down and transport of the products to other distant places whereas in weathering there is only break down of rocks in place. The surface of the earth is undergoing constant change, but it is so gradual as to be almost imperceptible.

There are so many natural agencies busy in shapping, modelling and sculpturing the earth so as to produce many types of scenery. The land surface varies from place to place, both according to the nature of the underlying rock and the agents modifying it. The chief agents of this process are running water, moving ice, ground water, winds, waves and current.

The work of all these agents consists of three aspects (1) break down or erosion of the rock material, (2) transportation of this material, and (3) deposition of the same material. In these aspects the transpartation is of prime importance and it consists of three processes, viz.

(i) The erosional agent dissolves the product and carries the material in solution.

(ii) The erosional agent carries the finest particles of sand and clayey mud with it as its suspended load.

(iii) The erosional agent drags and pushes the bigger fragments and pieces of rocks along its bed.

Thus the material transported in this way becomes the tool of the erosional agent while its energy depends upon the volume and velocity of a particular agency and nature of the tools and rock. When the erosional and transportational intensity of the agent is reduced, the depositional process starts and their gradual consolidation makes the sedimentary rocks, hence new lands are formed.

CONCEPT OF CYCLE OF EROSION

The concept of cycle of erosion has bean formulated by William Morris Davis (1850-1903 AD.) under the title of "Geographical Cycle of Erosion", in 1889. He had already propounded the concept of "Complete Cycle of River Life" in the year 1899. According to Davis landforms undergo sequential changes through time i.e. youth, maturity and old stage. The landforms are created by the internal forces of the earth, whereas the external forces level them down, till it becomes a featureless plain called as a peneplain. The whole period of the creation of relief features by endogenetic processes and their disintegration by exogenetic processes is called cycle of erosion. It has been defined by Davis as follows, "Geographical Cycle is the period of time during which an uplifted landmass undergoes its transformation by the process of land sculpture ending into a low featureless plain i.e. a peneplane".[39]

Davis was severely criticised by German geographers, particularly for using the term "Cycle" They were of the idea that this term was confusing and untenable. But instead of severe criticism this theory dominated the entire field of geomorphology from its inception in 1899 to 1950 throughout the world. It was basically concerned with the evolution of landforms in humid temperate areas but the cyclic concept was later on applied to all geomorphic processes e.g. arid, glacial, marine and karst cycle of erosion. Some geographers like Crickmay suggested modification in Davis geographical cycle but the basic idea remained the same.

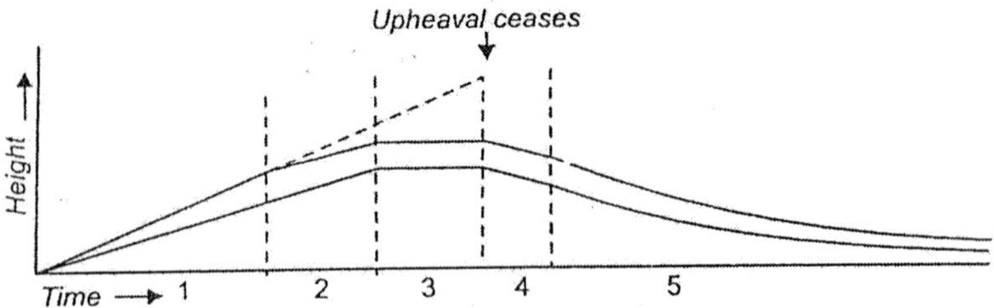

Fig. 1. Penck's Conception of a Cycle of Erosion during Long-continued Uplift. Upper and lower lines represent respectively ridge-crest and valley-bottoms.

CYCLE OF EROSION

The deposition or the upliftment of the landmass brings forth the different stages of erosional process which is known as cycle of erosion. In other-words the series of changes brought by erosion on a newly uplifted land surface from youth through maturity to old stage is known as cycle of erosion.

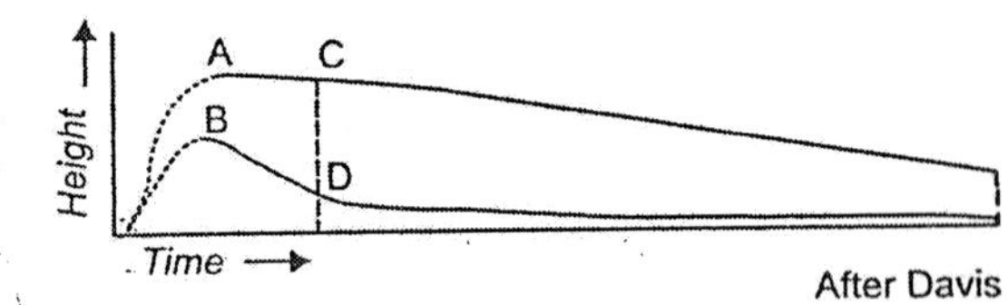

Fig. 2. Graphical Representation of an Erosion Cycle. AB, Initial relief; CD, Maximum relief.

(1) Youthful Stage

In this the river is in its upper course, so its work is primarily that of erosion or destruction. It is full of rapids and falls. The swiftness of the current, being strongest in the middle is constantly wearing away the bed of the river. The channel gradient is very steep which increases the velocity of the river flow, which instead increases its transporting capacity. The river carries big boulders of rocks alongwith it which help in "Valley deepening", through the vertical erosion. Hence the characteristic valley of the torrent course is steep-sided valley known as "ravine, Gorge or Canyon". The profile of the river in this stage is full of irregularities.

(2) Mature Stage

The mature stage comes in a river when it reaches in its middle course. The work of the river in this section is both constructive and destructive. The river is more powerful in cutting its sides or "Lateral Cutting", as the valley deepening is remarkably reduced. The summits of water divides are eroded i.e. there is a marked lowering of absolute relief. The lateral erosion leads to valley widening which transforms V-shaped valley into U-shaped. Because the rate of flow is slower, erosion is not so rapid, but the river bed is still being lowered by friction of rolling boulders. Some deposition takes place along the sides of river particularly during floods. A number of tributaries also join the main stream. The valley becomes broad and gentler in slope and attains its base-level.

(3) Old Stage

The river is said to be in old stage when it is flowing in its lower course. The river flows slowly, meandering in great loops over wide plains. Floods result in the widespread deposition of alluvium, e.g. the Ganga, Hwang-Ho and Mississippi. In the old stage the river becomes so sluggish in its movement that it is unable to transport the eroded material any further. So deposition is the rule in this stage. The river moves from side to side due to the hindrance of its own deposited material through a system of meanders. It lost its kinetic energy due to its extremely low channel gradient.

The valleys become almost flat with concave valley side slopes. The entire land escape is dominated by graded valley sides and divided crests, broad, open and gently sloping valleys having extensive flood plains, well developed meanders, residual monadnocks and extensive undulating plain of extremely low relief. Thus, the entire landscape is transformed into a flood plain or a peneplain, as revealed by the given fig. Same is the case with the wind, glacier and wave erosion as well, but the stages are not so clear as in river erosion. In these cycles the first stage is marked with picking and down cutting; the second with transportation while the third stage with deposition, as the intensity of the agent come to nullity.

MASS WASTING AND GRADATION

Mass Wasting

Mass movement, also called mass wasting, is the movement of the earth material downhill due to the pull of gravity. Once rock has been weathered, the pieces of rock that have broken away are capable of moving down along available slope. The quantity and speed of material transported by mass movement very with the type of earth material and the degree of slope. In some cases, whole mountain sides have given way and roared downhill in just a matter of movements. More often, earth material move grain by grain, only fractions of an inch at a time, even so, considerable quantities of material is transported over a long period of time. Removing of vegetation cover and water assists the movement of rock, besides gravity.

Landslides or Rapid Mass Movements

Mass movements are classified as either rapid or slow. Rapid mass movements are called landslides, which are further divided into rockslides, mudflows, earthflows, slumps, avalanches and rockfalls.

Rockslides

Often weak layers of clay, shale or limestone lie parallel to a slope or have joints and fractures parallel along it. The layers collapse and no longer support the rock layers above them thus causing the rocks to break and move downslope.

Mudflows

A mudflow is a mixture of rock, soil and water that flow downslope.

Earthflows

These are similar to that of mudflows. They slide over a bedrock surface as they move and are particularly common on steeper slopes within humid climatic regions.

Slumping

It is the downward and outward movement of rock and debris travelling as a unit or a series of units.

Avalanches

Landslides composed of ice and snow are called avalanches. These are generally confined to high mountain regions with steep slopes and abundant snow.

Rockfalls

It is the tumbling of individual pieces of rock downhill. These pieces accumulate at the base of a slope and build a considerable pile of rock debris called talus.

Slow Mass Movements

All forms of landslides can be easily detected by people. Slow mass movements are often more difficult to identify as they operate so slowly that they cannot be detected by the human eye. There are two types of slow movements (creep and solifluction).

Creep

It is the movement of soil and weathered rock down gentle slopes. There are no scars or breaks in the surface materials of slopes to suggest that creep has occurred or is occurring. It takes years together to find or detect the evidence of creep occurrence.

Solifluction

It is a feature of the higher latitudes where the sub-soil is permanently frozen, a condition called permafrost. In the summers water saturates the upper layers of the soil. Under saturated conditions, only a slight slope is enough to trigger the movement of the earth materials. The terrain of Tundra regions is especially modified by solifluction. They are capable of moving as much material as rapid mass movements are, but in a considerable period of time.

GRADATION

All the processes which tend to bring the surface of the lithosphere to a common level is known as the process of gradation.

It is done by the two processes degradation and aggradation. Degradation stands for levelling down the earth's elevated features and aggradation stands for filling up the depression. Gradation is thus a force which acts like an equalizer and tries to bring both the high and low lands to common grade or level.[40]

The large fragments of rock and big boulders are grounded to a powdery state, they rub against the bed of the river and against its sides. The result is that the river bed goes on being cut deeper and the sides go on becoming wider. This process is known as mechanical gradation. While many rock salts get dissolved in water as it flows over them in its course towards the sea. This activity is called as chemical gradation. A river does both the work of degradation and aggradation simultaneously.

The term gradation was used by Chamberlain and Salisbury in 1904. They recognized two categories of gradation processes, viz. (1) degradation and (2) aggradation.

(1) Degradation

Degradation means the levelling of the earth's surface by both destructional and constructional processes. It reduce the original elevation of the relief. River is the typical example of agency.

(2) Aggradation

Aggradation is the process by which the original irregularities of the earth's surface are removed and a level surface is created. For example running water, moving ice, waves and winds remove the products and carry them to other place, where it is deposited. It is a process of building up of channel.

MAN'S ROLE IN CHANGING THE FACE OF THE EARTH

Man is actively engaged in changing the face of the earth. Civilized man spreading everywhere and turning all parts of the earth's surface to his uses, has succeeded to some extent in reducing its physical differences. The earth is modified by human action is a natural fact of historical development. Reclamation of marshlands and submerged areas in Netherlands is cited as the greatest geographical transformation that man has brought about on the earth's surface.

According to Russel Smith, "Man is not merely a resident of this earth. He is builder and a geomorphic agent, an earth changer". His conquest of nature means that he has found in some measure how to utilize the forces of nature to his advantage and how to escape their detrimental influences.

Man more than any other animal leads a destructive life. The use of wood in construction and for fuel enables him to destroy the forests, so rapidly that all other animals are insignificant. The need for communication between distant parts of the earth has produced considerable changes in the configuration of coasts and its distribution of land and water. When land becomes valuable it is often profitable to reclaim ground from the sea. This is done along the flat coasts of most countries and to an unequalled extant in Holland, where most of the people actually live below sea level. The sea is kept out by a great system of artificial dykes and regulated sand-dunes.

The two landmasses having longest sea routes have been brought nearer by digging artificial channels, of which the Suez and Panama canals are the most remarkable examples. The rivers are interfered by drawing canals for irrigation and to prevent floods. Tunnels such as those through the Alps, through the Sahyadri Hills in Maharashtra between Mumbai and Puna and Jawahar Tunnel in Barrilal Kashmir etc. are some of the examples of man's roll in changing the face of the earth.

By diligent cultivation and careful selection the food grains have been produced from various species of wild grasses. In the same way many varieties of animals have been obtained by careful breeding which are specially fitted for the use of man. Without his interference they would have never existed. Useless creatures have been exterminated over wide areas and useful forms of life introduced in their place, sheep in Australia is the typical example. Useful plants of each continent have been transplanted wherever suitable conditions are found in the other continent. Potatoes, maize and tobacco brightens the fields of Europe while wheat, sugarcane and coffee spread over vast expanses of America.

Clearing land for cultivation, lumbering, changes in grasslands brought by the grazing animals, replanting trees, modification by fire and introducing new species are some of the examples of man's role in changing the face of the earth. Man as an agent of production transforms the surface of the earth for his own use by using different fertilizers, manure and irrigation etc. The relation between man and environment do not end at any stage. Human activity is often very closely integrated with the environment.

The above cited examples prove that there is a grand distinction between man and other creatures as he takes full advantage of physical features and his environment.

7
Fluvial Landscape

A landscape formed by the work of rivers during its erosion, transportation and deposition is known as fluvial landscape. The area or the region with in which the river carries on its functions is known as the drainage or the river basin.

THE DEVELOPMENT OF A RIVER SYSTEM

When rain falls or glacier melts or a spring spouts or a lake overflows upon the irregularities of a mountain peak, the water flows downward taking the shortest route. In this way little gutters are formed. These gutters when converge at a certain point give rise to streamlet. Many streamlets unite together to form a stream at still downward regions and several streams join to form a river, also known as running water. This running water forms a potent agent for general lowering of the earth's surface.

The source of a river is generally a mountain, where precipitation is heaviest and run off can flow down the slope. In other words it is called as catchment area of a river. The crest of the mountain is the divide or watershed from which the streamlets flow downwards, on both sides to begin their journey to the oceans. Several distinct stage like initial, youth, mature and old stage can be seen in the development of a river system.

(1) Initial Stage

Rivers generally take their origin from mountainous regions, owing to its considerable gradient and supply of water. Here the gradient is quite high, with the result the rives can perform a good deal of erosion. Deepening of the valley by down cutting of the river is dominant is this stage. In this stage the river passes through the rapids and falls. Deformation of landmasses is the rule of this stage.

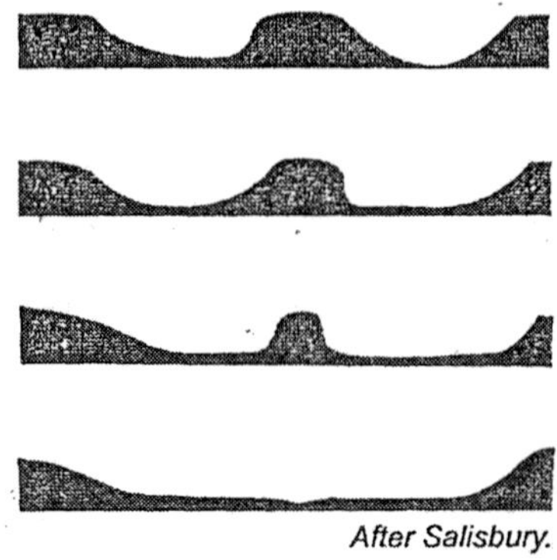

After Salisbury.

Fig. 1. Stages in the Destruction of an Interfluve by Lateral Corrasion.

(2) Youth Stage

In this stage the river's profile is irregular and the interfluves are undeveloped. Vertical cutting is most important, as a result the narrow valleys are formed due to the strongest current in the middle of the river. The valley is, therefore, deep and its sides are steep, known as gorges or canyons. As the lateral cutting is altogether absent is this stage, the divides are poorly developed.

(3) Mature Stage

In mature stage the erosional process is well advanced. The river starts lateral cutting or valley widening due to slow motion and a complex branching system of river is developed. The river has now come to the plain and is flowing over a more or less flat country. The drop in gradient decreases the velocity of the river as a consequence of which the erosive power is also diminished. The volume of water increases with the confluence of many tributaries and this increases the river's load. The transportation is predominant in this stage with some deposition. The river moves in a zigzag way giving rise to meanders and interlocking spurs.

(4) Old Stage

In this stage both the erosion and transportation has altogether stopped due to the sluggish movement of the river, thus deposition becomes the rule. The initial irregular surface becomes practically flat, which is called a peneplain.

The small mounds or hillocks of hard rock remain here and there in a peneplain which are known as monadnocks after the name of a hillock in New Hampshire (USA).

In this stage the river is frequently over flooded and builds up flood plains on both sides. The meandering of river give rise to ox-bow lakes and the deposition before joining the sea makes the delta. Both these are the characteristic features of a river in its old stage.

RIVER EROSION, TRANSPORTATION AND DEVELOPMENT OF VALLEYS

The whole works of fluvial processes or rivers are called three-fold work viz. erosion, transportation and deposition. The landforms which are shaped by these works are grouped under (1) erosional landforms such as gorges, canyons, broad and flat, mature senile valleys, multi-storeyed valleys, pot holes, rapids and waterfalls, structural benches, terraces, meanders etc. (2) depositional landforms such as alluvial fans, alluvial cones, natural levees, flood plains, terraces, deltas etc.

"Erosion, in general, is that process in which various erosive agents obtain and remove rock material or debris from the earth crust and transport them for long distance".[41] The erosional work of rivers depends on channel gradient, volume of water, velocity (Kinetic energy) of water, river load (its tools) and the nature of rock on which it flows. More steep the channel gradient, more will be the velocity of water and more the carrying capacity of the river load. If the velocity is doubled, the erosional power of the river increases four times. The magnitude of fluvial erosion mostly, depends upon the volume of water and the quantity and size of river's load. Because if the load is of fairly bigger size, it will roll down along the valley floor and will help in the deepening of the valley.

RIVER'S LOAD

The weathered rock material, eroded rock fragments, boulders and pebbles, which a river carries with it down the slope is called as its load. According to the way it is carried, it may be said to be of three distinct types.

(1) Material in Solution or Dissolved Load

The river water is able to dissolve certain types of rocks and mineral salts and hold them in solution known as dissolved load. Many rock salts get dissolved in water, as it flows over them in its course from the mountains to the sea. This activity is also known as chemical gradation. On an estimate about 50 tons of dissolved load is being carried off by river yearly on a sq. mile.

(2) Material in Suspension or Suspended Load

The finest particles of clayey mud, sand, silt, screes and talus are held suspended by water within itself while flowing downslope. With this material the river water usually appears muddy. It is called as dissolved load. It has been estimated that for every sq. ml. of earth's surface 200 tons of suspended material are being carried off by running water every year.[43]

(3) Material along the Base or Traction Load

This includes coarser materials, such as pebbles, stones, rocks and boulders which are dragged and rolled along the river bed. In the flow they collide with one another and are made into smaller angular pebbles. Due to constant friction with the river bed they result in pebbles, gravel and sand, the activity known as pebble grounding. The ability of a river to move the various grades of materials depends greatly upon the volume of the water, the velocity of the flow and lastly the size, shape and weight of the load. By doubling the velocity of a river, its transporting power is increased by more than ten times, that is why huge boulders are stranded and dragged along its bed.

TYPES OF RIVER EROSION

The erosional and transportational work of a river goes on simultaneously, comprising the following interacting processes.

(1) Corrosion or Abrasion

This is the mechanical grinding of the river's traction load against the banks and bed of the river. The rock fragments are hurled against the sides of the river and also roll along the bottom of the river. The corrosion takes place in two distinct ways.

(a) Lateral Corrosion

It is a side-ways erosion which widens the V-shaped valley.

(b) Vertical Corrosion

It is the downward action which leads to valley deepening.

(2) Corrosion or Solution

This is the chemical or solvent action of water on soluble rocks on which it flows, such as calcium carbonate (lime stone) is easily dissolved and removed in solution. According to Murray's estimate every cubic mile of river water contains about 762587 tons of dissolved mineral salts.[42]

(3) Hydraulic Action

This is the mechanical loosening and sweeping away of materials by the river water itself. Some of the water splashes against the river banks and surges into cracks and crevices. It helps in disintegration of rocks. The water also undermines the softer rocks with which it comes into contact. It picks up the loose rock boulders form it banks and transports them away. Hydraulic action is closely interrelated with chemical erosion and abrasion.

(4) Attrition

This is the wearing and tearing process of the transported materials themselves, when they are dragged and rolled along the river bed. The courser boulders are broken down into smaller stones, the stones into gravel and finally into sand.

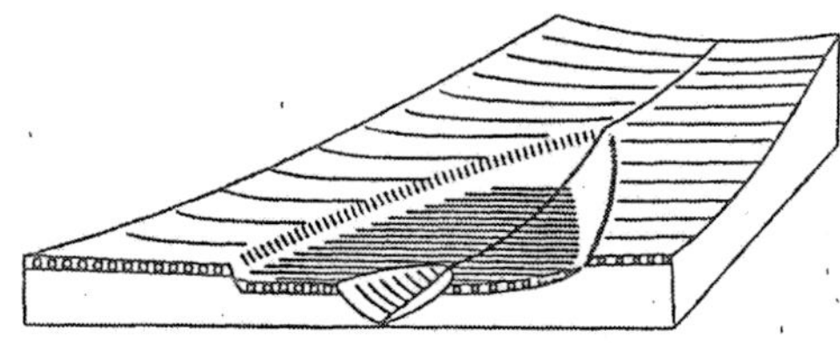

Fig. 2. Relation of Terraces of Composite River-profile.

EROSIONAL LANDFORMS OR DEVELOPMENT OF RIVER VALLEY

The river usually occupies a valley which it has either excavated itself or it was pre-existing. The latter type of valley is modified by river action. Development of a river valley depends upon several factors such as climate, composition of the bed rock and the surface relief.

The concentration water into the shortest route downhill causes the formation of little and narrow valleys often occupied by the rivulets. It may also be initiated by the presence of several pot holes more or less in a linear tract or by the retreat of a waterfall. As in the earlier stages of river development down-cutting preponderates over lateral cutting. The valleys are much more deepened and very less widened known as V-shaped valleys. These valleys are divided into two types viz. Gorges and Canyons.

(1) Gorges

A very narrow and deep opening of the river across a mountain having very steep sides is called a gorge. The gorges are formed due to active down cutting of the valleys through the mechanism of waterfalls and pot hole drilling in the youthful stage of fluvial cycle of erosion. The Indus gorge of Kashmir (J&K) is the example.

(2) Canyons

An extended form of a gorge is called as Canyon. The Canyons are very deep and narrow but long valleys, though the steepness of valley sides depend on the nature of the rocks. The sides may be very steep due to much resistant rock or may be undulating due to mixed (resistant and soft) rocks. The Grand Canyon of Colorado (USA) is the example.

RAPIDS AND WATERFALLS

A waterfall may be defined as a vertical drop of water of enormous volume from a great height in the longititudinal course of the rivers. Rapids are of much smaller dimension than waterfalls. These are liable to occur at any part of the river course, but they are most numerous in the mountain course where changes of gradient are more abrupt and frequent. The rapids and falls are formed due to a host of factors such as variation in the relative resistance of rocks, relative difference in relief, related rejuvenation, earth movements and fall in the level of the sea etc. The great force of a waterfall usually wears out a plunge-pool beneath. Waterfalls are formed by several ways viz.

(a) When water plunges down the edge of a plateau.

(b) When a bar of resistant rock lies transversely across a river valley.

(c) When a fault-line scrap caused by faulting lies across a river.

(d) When the glaciers produces hanging valleys.

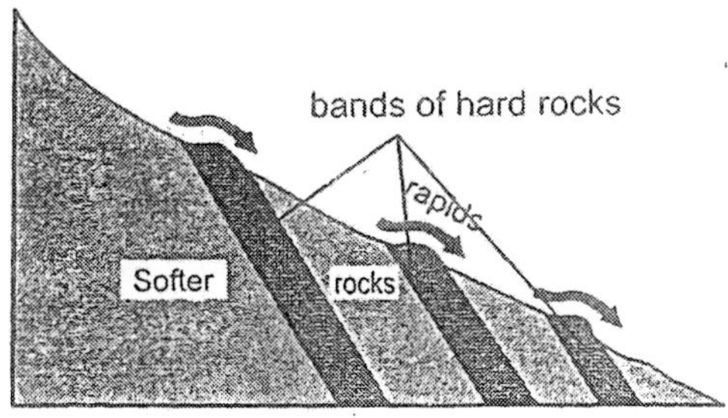

Fig. 3. Rapids, cataracts.

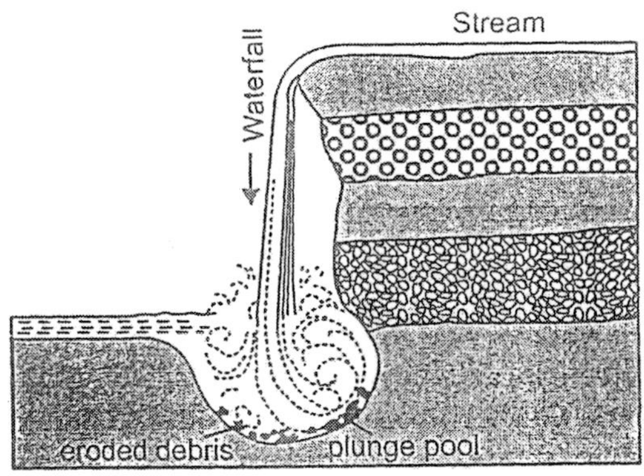

Fig. 4. A waterfall with plunge pool.

POT HOLES AND PLUNGE - POOLS

A hole worn in the solid rock, often at the foot of a waterfall, by the revolutions of stones and gravel, usually cylindrical in shape. These are generally formed in coarse-gained rocks like sandstone and granite. The fragments and gravel are kept in motion by the eddies of the swiftly flowing waters drill the pot-holes in like drilling machine.

The pot holes of much bigger size are called plunge pools. They are generally formed at the base of waterfalls due to the pounding of rocks by gushing water of the falls. Pot holes are the characteristic features of valley deepening.

Structural Benches

The benches or terraces formed due to differential erosion of alternate bands of resistant and soft rock beds are called structural benches. They are step like flat surfaces formed due to differential rate of erosion in the mountainous river coarse.

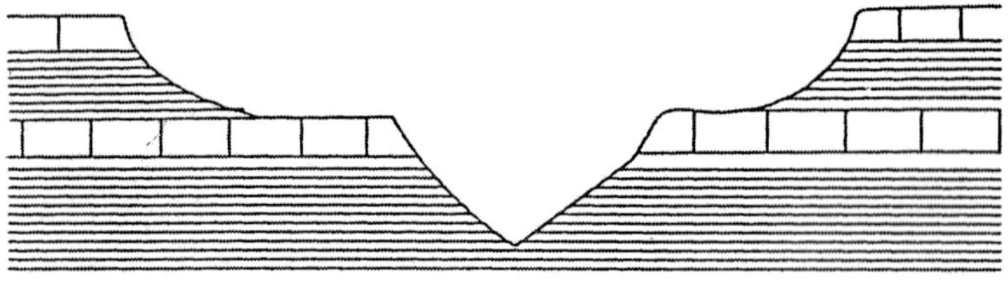

Fig. 5. Cross-profile of a Valley showing Structural Rock Benches.

River Terraces

The rivers form extensive flood plains during late mature stage and attain their graded curve of profile of equilibrium. The flood plains consist of thick deposits of gravel and alluvium. The rivers deepen their valleys due to accelerated rate of vertical erosion. Hence, the rivers form their new narrow valleys within the former flat valleys and thus terraces are formed on both the sides. In other words the river terraces are formed due to erosion of former flood plains consequent upon rejuvenation caused by either upliftment of the landmass or fall in sea level.

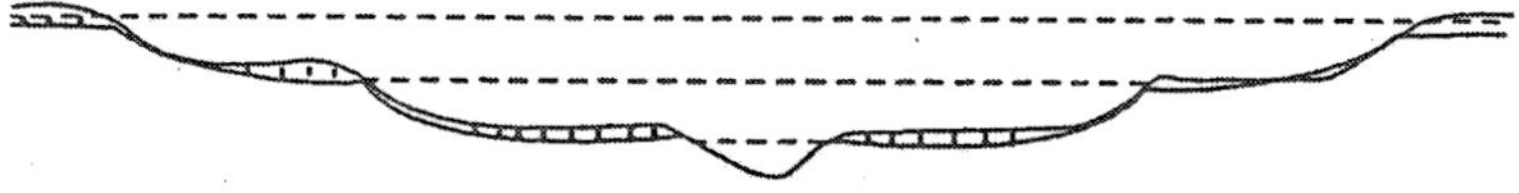

Fig. 6. Paired River Terraces.

Meanders

A curve in the course of a river which continuously swings from side to side in wide loops, as it progresses across flat country, is known as meander. The name has been derived from the river Meanderz of Asia Minor (Turkey), which in its lower reaches has a particularly twisting course. The meander is continuously begin accentuated by the river itself. On the concave side of a curve or bend the bank is worn away by the current, while on the convex side solid material is deposited. As a result the concave (outer) bank is often steep while the convex (inner) bank has a relatively gentle slope.

The concave side which is subjected to severe erosion results in the formation of vertical cliffs is known as cliff slope side. While the convex side which receives deposition of sand and gravel is characterized by gentle slope, is known as slip-off-slope.

A meander is usually semi-circular or S-Shaped channel of the stream. Meandering is most pronounced in the regions characterised by even surface and gentle slope, alluvial deposits and sufficient stream discharge. Meanders are the result of both erosion and deposition.

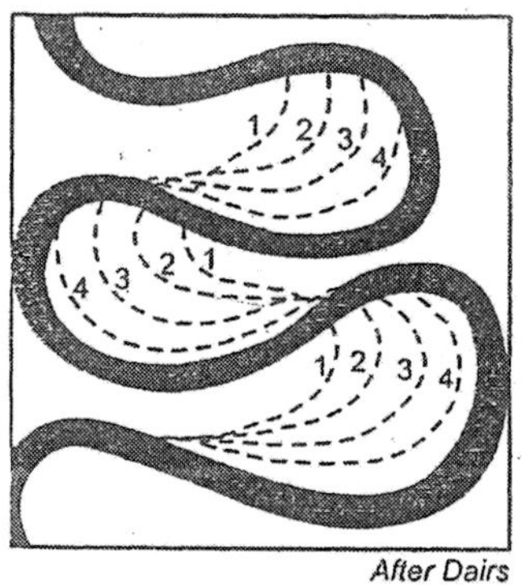

Fig. 7. Development and Downstream "Sweep" of Meanders.

Interlocking Spurs

As the stream flows on, the meanders migrate progressively outward with the interlocking spurs alternating with the undercut slopes. The meanders in the middle course are only the beginning of the downstream swing, for bends are restricted by the interlocking spurs. In the lower course the loops are enlarged across the level plain and meanders are fully developed.

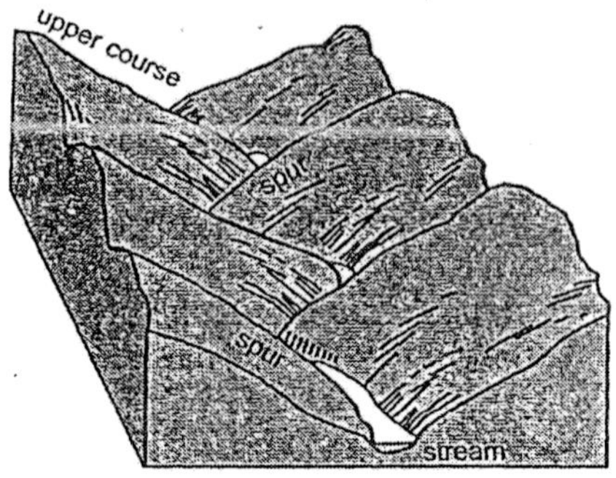

Fig. 8. Interlocking spurs

Ox-bow Lakes

A lake formed when a meandering river, having bent in almost a complete circle, cuts across the narrow neck of land between the two stretches and leaves a backwater. The mouth of this backwater is sealed by silt and mud till it is finally separated from the river and becomes an ox-bow lake or horse-shoe lake. In other words it is a detached loop of the meander. There is a frequent sedimentation of ox-bow lakes during floods, thus, they are converted into swamps. Here in the lower course of the river, due to slight or less gradient, deposition becomes the rule. Due to excessive deposition, the river's flow is continuously changing, which results in the formation of meanders and ox-bow lakes.

Peneplains

A peneplains is a region of low featureless plain having undulating surface. In the erosional work of a river almost all the elevated portions are worn down except more resistant rocks which stand above the general level of the land. These are, in fact, the end products of the normal cycle of erosion, known as monadnocks. When a peneplain is raised, it often becomes a plateau.[44]

River Transport

The river carries the eroded products with its current from the place of their existence to other places. Besides its own eroded material the river transports the other products of mass wasting processes. Materials from landslides, slumping, avalanches, underground water, wind-borne sediments are also largely transported by rivers. Of all the natural agents, the river is the most important and major agent of transportation. It transports the eroded material in suspension solution and traction. According to an estimate the Ganga carries 90 thousand tons, the Brahmaputra and Indus over 10 lakh tons of silt per day.[45]

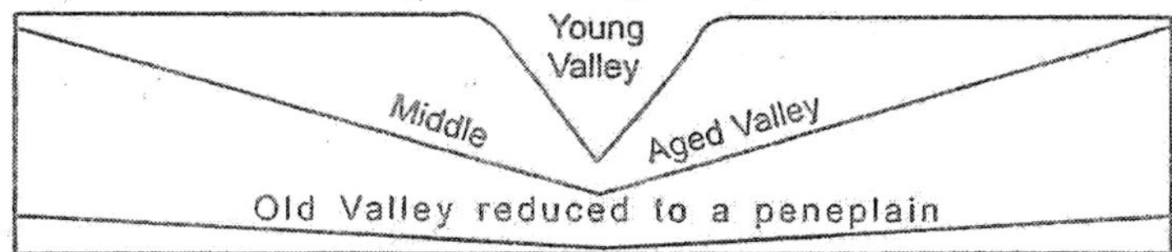

Fig. 9. How a River Valley is Broadened.

River Deposition

The deposition in the fluvial process starts when the river reaches in its last stage. There are various causes which are responsible for river deposition. They are as under:

(a) Decrease in channel gradient.

(b) Obstructions in the channel flow.

(c) Spreading of stream water over large area.

(d) Decrease in the velocity of river water.

(e) Decrease in the volume and discharge of water.

(f) Increase in the load of the river.

Where the velocity of the river is checked, deposition of the sediments takes place. Obstructions to the velocity are offered by the irregularities in the river course. In other words all the factors, which tend to diminish the velocity of rivers influence the deposition of sediments, loss of gradient being the main factor. Hence the plain stage of a river is mainly marked by deposition.

Loss of volume resulting in the decrease in velocity also causes deposition. This is specially manifested after floods. During flood time a river does considerable damage by means of its sudden increase in volume but when the flood sub-sides, the volume of the river gets diminished and consequently there occurs a good deal of deposition.

DEPOSITIONAL LANDFORMS

As the rivers flow down towards the sea, they go on aggrading their courses, their banks and the plains on their left and right. In its last stage the river becomes so sluggish that deposition becomes the rule. It results in various characteristic forms of dunes, spurs, bars, embankments, fans and deltas.

Dunes

A mound or ridge of sand formed, either in the desert or along the sea coast, through transportation. The sand particles are carried and piled into a heap, which gradually increases in size till it becomes a mound or ridge.

Spurs

It is a kind of projection of higher ground over the lower one and as such is quite reverse of a valley. On the map a spur is shown by contours running out from the main features.

Bar

A ridge of sand shingle formed in the sea coast across the mouth of a river or the entrance to a bay or harbour and laying approximately parallel to the coast. It is also called an offshore bar.

EMBANKMENTS OR LEVEES

As the meanders sweep own stream, they make river aggrade coasser material near the banks. This deposit results in the formation of levees or embankments. These levees rise on the banks of the river and the slope is 5 to 10 feet per mile. In other words the narrow belt of ridges of low height built by the deposition of sediments by the spill water of the stream on its either bank is called levee or embankment. They are formed by deposition during flood periods parallel to the river valleys. Sometimes, natural levees are also used for agricultural purposes because water table is very high.

ALLUVIAL FANS AND ALLUVIAL CONES

Alluvial fans and cones are formed by deposition near the foot of the hills where the river enters the plains. With the decrease in the gradient, the transporting capacity of rivers diminishes. Consequently, coarse gravel, pebbles and sand are deposited in conical heaps known as alluvial cones. Several cones, when combine give a fan-like structure known as alluvial fans.

The shape of alluvial fans is usually semi-circular the appex of which is located at the mouth of narrow opening through which the stream comes out of the hills. Its longitudinal profile is concave at its appex, while the transverse profile is convex. The size of alluvial fans varies from a few km to several hundred km. The slopes of fans are much gentler than those of alluvial cones. The cones are made of coarser materials whereas the fans are of finer sands.

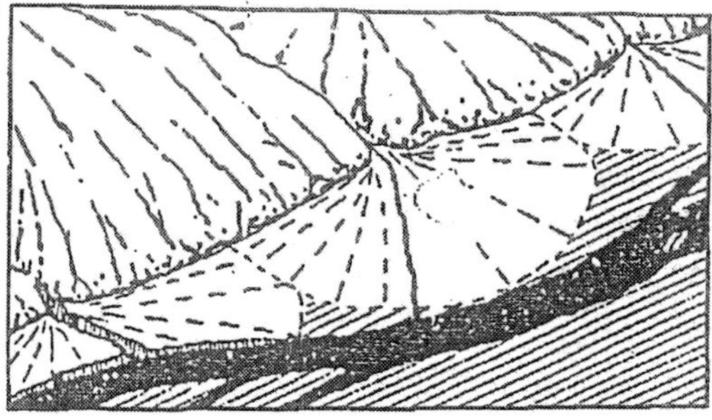

Fig 10. Alluvial Cones.

DELTAS

At the last stage the river falls into the sea or a lake, depositing its finest material in the form of alluvial tract at its mouth. In simple words deltas are triangular structures built by river deposits. The word delta has been derived from the Greek's fourth letter of alphabet (Δ). It was first used by Greek historian Herodotous (485-425 BC) for the triangular depositional feature at the mouth of river Nile (Egypt).

The formation of delta starts with the regular sedimentation at the mouth of the river, on the sides of the stream channel, in the bed of the river. Thus, an extensive fan is formed which slopes towards the sea. These fans are coalesced and a delta is formed. The deposits obstruct the free flow of main river and hence it is divided into several branches. The alluvial treat is, infact, a seaward extension of the flood-plain. Due to the obstruction caused by the deposited alluvium, the river may discharge its water through several channels called distributaries.

Deltas differ much in their size, shape, growth and importance. For instance, the Ganga delta is almost as big as the whole of west Malaysia. A number of factors such as the rate of sedimentation, the depth of the river, the sea bed, tides, currents and waves greatly influence the eventual formation of deltas.

Conditions Favourable for Delta Formation

(a) Active erosion of the river in its upper course to provide extensive gravel, sand and silt to be eventually deposited as deltas.

(b) The coast should be sheltered preferably tideless.

(c) There should be no strong current running at right angle to the river mouth, as it can wash away the sediments.

(d) Any large lake in the way or river course can filter off the sediments, thus unfavourable for delta formation.

(e) The sea should be shallow adjoining the delta as the sediments will disappear in the deep waters of the sea.

RIVER REJUVENATION

The process by which erosive activity is renewed, applied especially to rivers which are said to be rejuvenated. The rejuvenation of a river may be brought about by the uplift of the land over which it flows or by a fall in sea level, so that it begins to cut into its bed once more and a new cycle of erosion begins.

Rejuvenation takes place at a late stage in the history of a region. The region may be uplifted so that the downward corrosion may be renewed. It occurs when the river has attained full maturity. But rejuvenation can take place at any stage of river development. The first cycle of erosion is replaced by second one, known as successive cycle of erosion. In other words, a particular landform, while udergoing the first cycle of erosion may be disturbed so as to undergo a few other cycles successively. The two major factors on account of which the rejuvenation takes place are as follows:

(1) Negative Movement

It occurs when there is an uplift of land or a fall in sea-level. It steepens the gradient and as a result down-cutting is renewed. The river also with its renewed vigour cuts into the former flood plain, leaving behind terraces on both sides of the river. The point of break in the graded profile of river (where old and rejuvenated profile meet) is known as rejuvenated head or knick point.

(2) A Positive Movement

It occurs when there is a depression of land or a rise in sea level. It will submerge the lands along the coast. The flow of the river will be checked and large quantities of river load will be dropped. The lower course of the river may be partly in the sea and features of deposition are shifted upwards to the middle course. If there is a rise in the sea level, the mountainous course is very little effected.

As and when the new cycle of erosion starts the following changes begin to take place.

(a) The rivers begin grading their course and forming a new track of their erosion.

(b) The rivers begin incising their course below the already existing floodplains, resulting in the formation of alluvial terraces on both sides.

(c) As down cutting starts, the meanders begin to change their form and shift in their positions. Accordingly as the incision is more rapid or less, intrenched or in grown meanders are formed.

8
Aeolian Landscape

Although wind is the least effective of the major erosion agents, it is capable of modifying existing landforms. Aeolian processes and landforms are most obvious in arid regions (deserts), where lack of vegetation not only makes wind erosion, transportation and deposition possible but enables the resulting landforms to be observed. The deserts defined as barren, desolate and plantless areas are of two types viz. hot and cold deserts. Aeolian processes are not active in cold deserts as its surfaces are always covered with permanent ice sheets (glaciers). Thus, wind is the chief agent of denudation in the hot and temperate deserts (arid regions) of the world. Arid zones cover about one third of the earth's surface.

The chief factors that limit the wind action are rainfall shortage, dirth or lack of vegetation and excessive evaporation over precipitation. In the absence of roots that bind the loose earth particles and the humidity of the soil that keeps it from being blown off, the wind attains special force of erosion in arid regions of the world. These include the Sahara desert of N. Africa, the Arabian desert of Middle East, the Victoria desert of Australia, the Kalahari desert of S.W. Africa, the Sonora desert of N. Wn. Mexico, Sn. Arizona and California, the Atacama desert of Peru and Chile, the Afghanistan desert, the Baluchistan desert and the Thar desert of India and Pakistan.

THE MECHANISM OF WIND EROSION

Arid landscape is the result of many combined factors, one reacting upon the other. Meagre rainfall (less than 5 inches per annum), coupled with very high temperatures (about 87 **degree** F) and a rapid rate of evaporation are its principal causes. Weathering is the potent factor in reducing rocks to sand in arid regions. Some of the moisture penetrate into the rocks and sets up chemical reactions in the various minerals. Secondly, the intense heating during the day time and rapid cooling at night by radiation, causes expansion and contraction respectively. It finally set up stresses in the rocks so that they eventually crack. Again the outer surface of the rocks is heated by the sun, while the inner rock remains quite cool. It causes the outer surface to expand and so rise itself off from the interior. Hence it peels off in successive layers with the passage of time. This onion-peeling process of mechanical weathering is called as exfoliation.

Wind erosion has three fold activity (a) deflation, which is a Latin word deflare, meaning "to blow away". This term was first used by Richthofen, a German geographer. (b) Abrasion or mechanical erosion performed by wind with the help of sand grains, which become its tools. (c) Attrition or the wearing and further disintegration of tools.

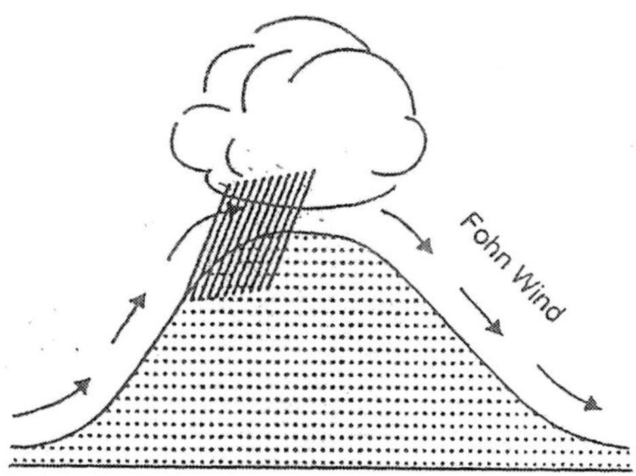

Fig. 1. Diagram illustrating the formation of the Fohn Wind.

(a) Deflation

This involves the lifting and blowing away of fine dust and loose materials from the ground. The sands and pebbles may be carried in the air or rolled along the ground depending on the size of the grain and velocity of the wind. The finer particles of sand and dust are removed severed miles away from their original place and are deposited even outside the desert margins. Through this process, the altitude of deserts is lowered to form large depressions called deflation hollows. Qattara depression 450 feet below sea level in Sahara Desert is a typical example of such hollows.

In coastal regions particles of the beach sand are dried during recession of the sea water at low tides to which wind carries inland. It has been calculated that thousands and thousands of tons of such sand and dust particles are carried annually to great distances. When wind blows a huge quantity of the dust particles is lifted up and this covers the sky creating a suffocating atmosphere. The particles of sand and quartz mixed in the wind are able to polish or etch the surface of rock besides produce flutes, channels and carve out hollows.

(b) Abrasion

The process of wearing away of part of the earth's surface by the action of wind utilizing moving debris e.g. sand, as an abrasive material is called abrasion. The impact of such blasting results in rock surfaces being scratches, polished and worn away. This process is most effective near the base of rocks, where the amount of material, the wind is able to carry, is greatest.

The wind abrasion depends largely upon the character of the blown rock particles and the bed rock. Generally the effect will be at its maximum when (1) the blown particles are hard, (2) the rock bed is soft, and (3) the velocity of wind is great. Thus, the rocks consisting of both hard and soft parts get differential abrasion. The wind abrasion decreases drastically with increasing height from the ground. It becomes inactive beyond about 2 meters height, thus only undercuts the upstanding objects from all sides. In this way several desert features are produced by abrasion such as, mushroom rocks, table-rocks, broad shallow caves and pedestals etc.

(c) Attrition

The mutual wearing and tearing down of rock particles during its transport by wind, so that they become reduced in size, smooth and rounded. The sand particles not only abrade the exposed bed rocks but are also abraded themselves by colliding against each other. The result is that the individual fragments acquire a rounded shape. It is all because of the fact that while on transit, the sand particles get abraded equally almost on all sides. The other contributing factors are the greater velocity of the wind and greater length of its transit. The rounded sand of the deserts is usually called millet seed sands because of its resemblance with millets. This roundness of the individual fragment can easily distinguish a wind formed sandstone from a river born sandstone as in the latter case the grains will not be perfectly spherical.[46]

MAJOR LANDFORMS OF WIND EROSION

The continued erosional work of wind through the mechanism of deflation, abrasion and attrition produce various characteristic landforms in the desert areas. These features are blow-outs or deflation hollows, desert pavement, mushroom or pedestal rocks, dreikanter, stone lattice, zeugens, yardangs and inselbergs etc.

(a) Blow Out

The chief landform produced by wind action is the blowout or deflation hollow or basin. It is a hollow formed in light soil or in sand, e.g. in a dune, by deflation. Its formation may be initiated by destruction of the natural vegetation, or by overgrazing by animals, exposing loose, finally divided material easily lifted by wind. Rain fill this depression creating a pond or oasis in the desert areas.

(b) Desert Pavement

A relatively smooth mosaic-like area in a desert region, consisting of pebbles closely packed together after the removal of finer material.[47]

(c) Mushroom Topography

It consists of tabular mass resting on a thin pillar connected with the pedestal. This type of rock masses in desert areas look like mushrooms. These features come into existence by erosive action of wind through etching, fluting and undercutting the rock masses. They are also known as polizfelson.

(d) Dreikanter

The carving stones faceted in triangular shape is known as dreikanter. Such stones are generally pitted in the direction from which the wind comes. By wind erosion sometimes, as many as three edges are produced on the pebble.

(e) Stone Lattice

In the desert topography stone lattice is a common feature produced by wind erosion. The rock when abraded by powerful wind becomes uneven as only the weaker section are removed. Such pitted and fluted rock surfaces are called as stone lattice.

(f) Wind Bridges and Windows

Due to continuous abrasion on stone lattice the holes are formed, by powerful wind. These holes are gradually widened and reach the opposite side of the rocks known as wind windows. They are further widened through the processes of abrasion and deflation in such a way that an arch like features having intact roof are formed known are wind bridges.

(g) Zeugens

In mushroom topography sometimes the pedestal (pillers) of softer rocks are completely removed by undercutting so that the whole mass is overturned or turned turtle. The feature thus produced is known as zeugen.

(h) Yardangs

The zeugens are further modified by the continuous wind erosion, so that the whole tabular mass becomes irregular in form. There are the steep sided undercut overhanging ridges of rock, separated by long corridors which have been cut out by wind borne sand. They are roughly parallel and lie in the direction of the dominant wind.

(i) Inselbergs

An isolated hill in a hot dry region, inselberg, a German word, first used by Passarge (1904) to indicate sharply rising residual hill above the flat surfaces in South Africa. Such hills or mounds of relatively resistant rocks are also called "bornhardts" after the name of famous geomorphologist. It may sometime rise to well over 1000 feet above the surrounding plain. sometimes these hills are pyramid-like with a cap of hard rock and sometimes they are dome-shaped having been made out of granite rocks. They make an appearance of the island in the desert.

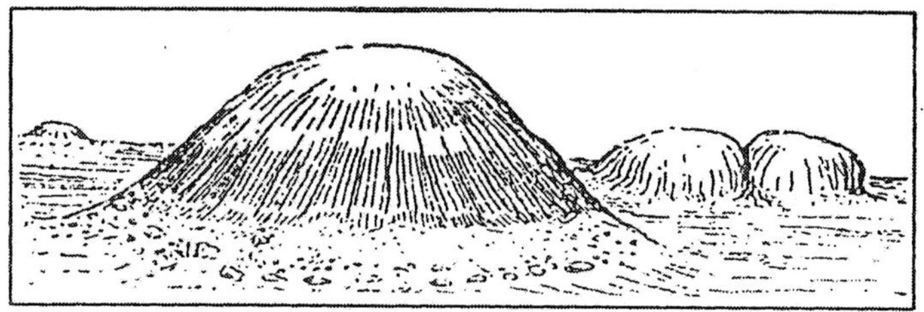

[From photographs by A.D.N. Bain

Fig. 2. Granite Domes (Inselberge) near Lemme, Nigeria.

TRANSPORTATION

The transportational work of wind totally differs from that of river and glacier etc. The wind transports the materials through the mechanism of suspension, saltation and traction. The fine material (dia.O.2mm). is kept in suspension by upward movement of air which is known as dust. While the very fine material is called as smoke or haze. The larger particles and fragments of rock are rolled along the bed rock or transported through bouncing, leaping and jumping. The latter process of transportation is čalled saltation, whereas the former mechanism of transport through rolling is called surface creep or traction. In both these processes the load or material has direct contact with the bed rock. By this mutual contact and constant friction both are ground down into fine sand and smoothened surface.

MAIN CHARACTERISTICS

(1) Wind transport involves larger areas and greater distances. It is elstimated that in the Wn. U.S.A. wind transports 850 million tons of sand to over a distance of 1440 miles.

(2) The direction of wind transport is variable as it often changes its direction.

(3) Only very five material is transported to great distances in one step whereas the coarser material is transported

in steps and stages by rolling, leaping and jumping.

(4) The wind transports the material at the ground surface and above the ground i.e. in air.

DEPOSITION

When the forward movement of wind is arrested the load is at once deposited. As a matter of fact deposition starts when the strength of wind declines or some kind of obstacle comes in its way. These obstacles may be bushes, forests, marshes, swamps, lakes, rivers and walls. When the wind carries the dust particles to regions outside the desert margins and deposit it is in accumulations of huge size is called loess. The coarser materials deposited in hillocks within the territories of desert are known as sand-dunes.

Such wind formed deposits are called aeolian deposits after the name of 'Aeolus', the ancient Greek god of wind. One peculiarity of these deposits is the sorting action i.e. the deposits are arranged according to their size and weight. The finest particles are carried farthest than the heavier and the larger ones.

DEPOSITIONAL LANDFORMS

The following are some of the major features of wind deposition.

(1) Loess

A deposit of fine silt or dust which is generally held to have been transported to its present situation by the wind is called loess. In other words the fine dust blown beyond the desert limits is deposited on neighbouring lands as loess. It is yellow friable material and is usually very fertile. It is a fine loam, rich in lime and extremely porous. The rain sinks quickly into it and its surface is normally dry, but when irrigated it is very fertile. Through the loess, streams have cut deep valleys, often with vertical walls, known as badland topography.

There are extensive deposits of loess in central Europe, central United States and fertile Black Earth of U.S.S.R. It is also spread over Hwang-Ho basin in NW China. According to an estimate the loess deposits cover an area of 250 thousand sq. miles with a depth of about 200 to 500 feet.

(2) Dunes

A mound or ridge of sand formed in desert through transportation by the wind. The sand particles are carried along by the wind and piled into a heap, which gradually increases in size till it becomes a mound or ridge. The dune is often commenced where an obstacle of some kind exists and the sand is heaped against this until it is covered and the sand falls over on the leeward side.

There are two types of dune, the crescent-shaped dune or "barkhan" and the long, narrow ridge-like dune or "self-dune", extending in the direction of prevailing wind.

These two dunes will be described in more details as under:

(a) Barkhan

It is a crescent or sickle or half-moon shaped dune with two tapering arms. These transverse dunes formed in groups or individual depending on the obstacle. The word barkhan or barchan has been derived from a Turkish word. They are most prevelent in the deserts of Turkistan and Sahara. They occur transversely to the wind so that their horns thin out and become lower in the direction of wind. The windward side is convex, while the leeward side is concave. The rate of advancement varies from 25 feet a year for the high dunes measuring upto 100 feet high per year.

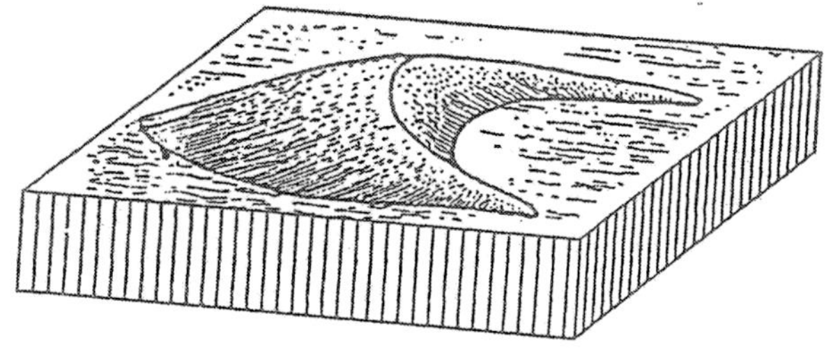

Fig. 3. Barchan.

(b) Seif-dune

These are long narrow ridges of sand, often over a hundred miles long lying parallel to the direction of prevailing winds. The word "seif" has been derived from Arabic meaning sword. These longitudinal dunes are formed by the modification in the shape of the transverse sand dunes. If the wind is faster and moves violently, it carves out a channel through the minor hollows and concavities on the lee side. The sand is trailed out along them and then and the seif dune is divided into two separate ridges parallel to the prevailing wind direction. Extensive seif dunes are found in the Sahara desert, south of Qattara depression, the Thar desert and the W. Australian desert.

LANDFORMS DUE TO WATER ACTION

As a matter of fact, the deserts are entirely without rain, as the annual precipitation is about 5 to 10 inches and often comes in irregular showers. But sometimes the rain falls in torrential downpours as the 'thunderstorms' do occur in arid regions. These rains bring devastating effects, drowning people who camp in dry deserts streams and often flooding mud-baked houses in the pases. Large quantities of loose rock material (gravel and sand) is transported in the sudden raging torrents or 'flash floods', down the hill side. They cut deep 'gullies' and ravines forming 'badland topography'. There is so much material in the flash floods that the flow becomes liquid mud. When the masses of debris are deposited at the mouth of the valley an alluvial cone or fan is formed. The water flows through several channels carved over this dry delta.

The gullies are further deepened by vertical corrosion by raging torrents during the occasional cloudbursts. These valleys are known as wadis, which are dry for most of the time. Sometimes water collected in a depression or a desert basin does not completely disappear by evaporation or seepage and a temporary lake is formed. Such lakes contain a high percentage of salts, because of high evaporation. In United States and Mexico these lakes are known as 'playas' or 'salinas' or 'salars' and are called 'shotts' in N. Africa.[48]

9
Glacial Landscape

GLACIARS

In the past ice age, million of years before, almost all the continents of the world were under ice-sheets. These glaciers have left their marks, which are still visible on the southern portions of each continent. They eroded the landscape and formed new landforms by erosion and deposition. The shapes of many of our present landforms are indeed the direct result of those previous glacial activity. However, today glaciers cover only about 10% of the earth's surface. At present they are confined to higher latitudes and over altitudes (mountains), where temperatures always remain below freezing point.

The study of glacier requires first a thorough understanding of the phenomenon of snowfall, snowfield, snowline and difference between snow and ice. Snowfalls in place of rain, when water vapour condense at a temperature below freezing point and forms minute spicules of ice, which unite into crystals. These crystals aggregate into 'snow flakes', a dry, powdery, light and white feathery stuff. When the snow accumulates layer after layer, it is pressed under the weight of upper layers and thus the lower layers transform into a crystalline compact form known as 'ice'. The places or regions of permanent snow for the whole year are called as 'snow fields' and the line above which there is perpetual snow is known as 'snow-line'. The snowline is higher on steep slopes than on gentle slopes. Sufficient snowfall, cool temperatures, a low rate of melting in summer and less evaporation ensures the growth of snowfield.

GLACIAL FLOW

A glacier (moving ice) is a mass of ice and snow that moves slowly down the valley slopes towards the sea under its own weight, slope and force of gravity. What causes a glacial surge is still unknown, but it is thought to be related to the accumulation of melt - water under the glacier which assists in lubricating the path over which the glacier flow. Long range studies have shown that glaciers advance more rapidly in their centers than along their edges because friction is reduced in that area.[49] Its life and continuity depends upon the snowfield and wastage in the way. The movement of glacier starts when the weight of the mass becomes more than the retarding friction along the slopes and is also dependent on the temperature. Its movement is also serpentine and conforming to the trends of the valley in which it flows. Ice thus behaves as viscous mass and solid crystalline which flows by the application of sufficient force coming from overlying mass of ne 've' and snow. The glaciers near the sea sometimes project towards it and parts of it break there, which float on the sea water. This floating ice hill is called as iceberg. These are very dangerous for navigators as they destroy ships.

As for as the movement of a glacier is concerned it varies from one inch to many feet in different glaciers. The velocity of it increases with the increase in slope of its bed, thickness of its ice, straightness of its course and the temperature. Glacier movement also varies with increased depth in the glacier. Two zones of movement are identified. The lower zone, called the 'zone of flow', consists of glacial ice. The lowest layer, in this zone are under great pressure and partially melt into a plastic like consistency. The zone of flow moves at variable rates over objects in its path. The upper zone of the glacier, called the 'zone of fracture', extends to a depth of 100 to 200 feet. It consists of brittle glacial ice, firn and snow. Unlike the zone of flow, the zone of fracture will not bend as it flows over objects. Instead, this zone cracks apart whenever the glacier bends, forming large breaks at the glacier surface called 'crevasses'. Some of these may exceed depths of 50 feet.

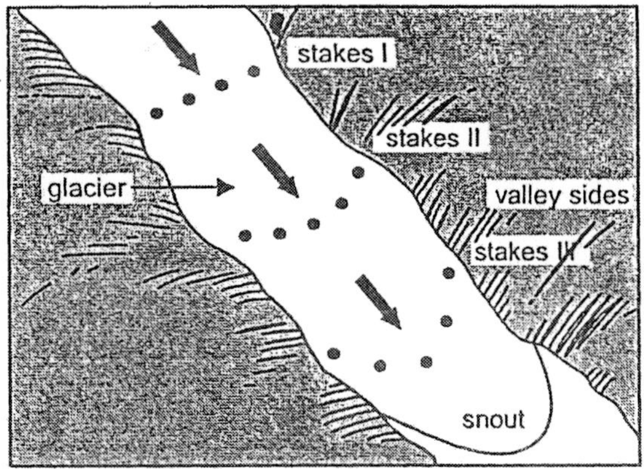

Fig. 1. The different rate of glacial movement. A glacier moves faster in the centre than the sides.

TYPES OF GLACIERS

Depending upon their structure and location glaciers are classified into three types which are as follows:

(1) Valley Glaciers

The glaciers that flow down valleys like rivers are called valley glacier or mountain glacier or Alpine glacier. They are always confined to the valley walls and are commonest type. These are found in Rockies, Andes, Alps, Himalayas and Norway etc. Those of Karakoram and Alaska are 65 to 129 km in length and glaciers occupying hanging valleys are largest in the world. The shape of the valley glaciers varies from mountain to mountain as is clear from the following descriptions.

(a) Cliff or Cornice Glacier

These are the glaciers which are found on the face of the cliff and are of various shapes. Sometimes it is circular or semicircular, at other time it is linear or irregular.

(b) Cascade Glacier

These are the glaciers which end on the face of a cliff and consequently assume a cascade - like shape having broken ends.

(c) Recemented or Reconstrued Glacier

On the termination of a glacier, on the face of a cliff, ice blocks are broken and scattered. These may get accumulated and recemented to form an ice mass known as reconstrued glacier.

(d) Rock Glacier

Rock glaciers are quite common in Alaska. If the glacier's surface is covered with angular rock fragments throughout its journey, it is called as rock glacier.

According to their expanse and form they may be classified into six types viz. (i) expanded foot type (Toku in Alaska); (ii) dendritic type (Baltaro in Himalayas and Tasman in Newzealand; (iii) readiating type (Alps in Europe); (iv) horse-shoe type (Canada in USA); (v) inherited basin glaciers; and (vi) spitzbergen glacier.

(2) Piedmont Glacier

When two or more valley glaciers come out of the valley and fall to the plain they unite to form a bigger ice mass which is called a Piedmont glacier. These are intermediate in form as well as origin between the valley glaciers on the one hand and ice sheets on the other. It is a type of glacier formed at the foot of a mountain. Melaspina and Bering glaciers of Alaska (USA) are the typical example of Piedmont glaciers. Such glaciers are found only in colder areas and not in the tropical or temperate regions because they melt when they reach foothill zone.

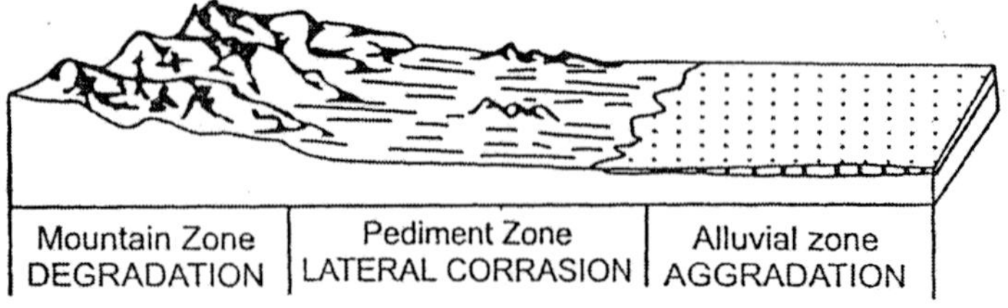

Fig. 2. Johnson' Theory of Pediment Formation.

(3) Ice Sheets

These are huge covers of ice enormous in size and thickness, that rise high above the mountain peaks and completely bury them. Ice sheets may be divided into the following two sections.

(a) Continental Ice Sheets

Vast snow covered regions, limitless in expanse, are the Ice sheets. Greenland and Antarctica provide the only examples of continental ice sheets at present. The Greenland ice sheet covers half a million sq. miles is enclosed within a mountainous rim and rests over a rock floor which is only 1600 mts. a.s.l. In the middle, ice is 8000 ft. thick, which has been estimated by seismology, measured by snowgauge.

(b) Ice Cap

They are much smaller than ice sheets covering only hundreds of sq. mts. and are sometimes called plateau glaciers. At present they are met within Iceland. Scandinavia, etc. In some cases they radiate the valley glaciers.

WORK OF GLACIERS

The year 1837 marks the date since when some study about the work of glaciers has been made. There are three contrasting view points as regards glacial erosion. One school of thought believes that the glaciers protect the crust as it covers the bedrock. Prof. Heim is the supporter of this thought who has propounded his theory in 1885. Another school of thought regards the glaciers as active potent agents of erosion and deposition. Dr. Hess has greatly exaggerated the erosional power of the glaciers and has brought them at par with the work of running water. The modern thought or third school of thought propounds that, so long as the glacier is stable, it is a protector, but when moves, it does the work of erosion, transportation and deposition.

MECHANISM OF GLACIAL EROSION

The erosional work of glaciers is accomplished through the mechanism of abrasion, plucking and polishing. Geomorphologically the pure ice mass is inactive but when it helds and carries coarse debris at its base it becomes an active agent of erosion. Thus a glacier uses this coarse material as tools for its erosional work through abrasion. While the large boulders of rock fixed in glacier base are detached from the base rock by the mechanism of plucking. Thereby big grooves and hollows are formed. The pieces of rock held by the glacier are dragged along the floor so that both the rock basement and the boulders are ground smooth by mutual contact. It results in planning, polishing and scratching the boulders as well as the floor of the valley. This work of glacier is comparable to the work of a lapidiary which grinds a precious stone by fixing it in a matrix till its irregularities are smoothened. This flatened floor of the valley is known as the "glacier pavement", which is full of scratches and streaks indicating the direction of the ice movement.[50]

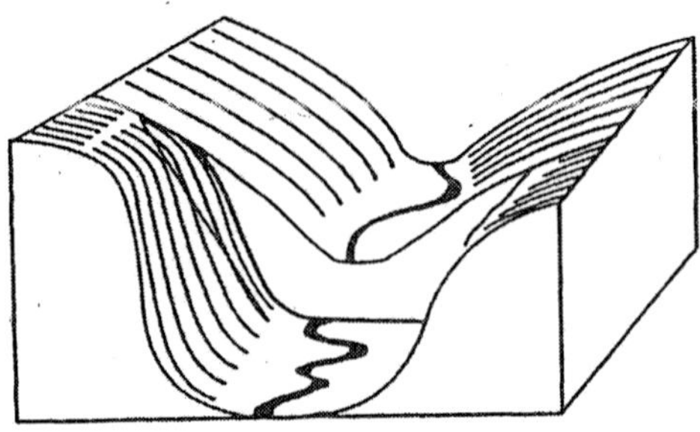

Fig. 3. Contrast between Water- and Ice-eroded Valleys.

The glaciation or glacial erosion depends upon several factors, some of which are as under:

(a) Thickness of Glacier

The more the thickness, the more the glaciation is the rule, thus the thickness of a glacier is very important factor in glacial erosion.

(b) Amount of Rock Material

More the material griped by the floor of a glacier, more will be the abrasion, plucking and polishing of the bed rock. The firmness exercised by glaciers on the stones is an important factor in glacial erosion. If this grip is loose by melting of ice, the erosion is not pronounced.

(c) The Velocity of Glacier

The greater the velocity of the glacier, the more is the erosion. Thus steeply inclined surfaces help speedy glaciation, while the slowly moving ice sheets are less powerful agents of erosion.

(d) The Nature of Bed Rock

The weak bed rocks are easily eroded away instead of hard rock which take a considerable time to be eroded away. Thus the glacial erosion is more pronounced in the weaker bed rocks than the hard rocks. One peculiarity of glacial erosion is that it can erode the valleys below the sea level. On the contrary, the down cutting by rivers stop at the sea level. This activity of glacier is known as 'over deepening'.

GLACIAL TRANSPORTATION

The glaciers transport enormous bulk of load from the higher latitudes and over altitudes to downhill and sea coasts. All the products of erosion are in a state of progressive transits outward and downhill by the glacier ice. A huge quantity or rock debris is carried by glacier on the upper surface, which is called the 'glacial drift'. The rock debris comes from the valley sides. Avalanches also contribute to the rock debris. By the process of plucking some rock material is picked up from the bedrock. Regulation i.e. melting under pressure and subsequent freezing, is a very important process in this respect. The materials are arranged along the sides, under the glacier and at the terminus. All the rock materials tend to accumulate at the terminus. From these terminus the melt water in different types of glacial streams carries the rock material to the seas and lakes. In coastal regions glaciers directly carry their rock material (debris) to the sea.

MAJOR EROSIONAL LANDFORMS

There are certain erosional landforms which may be said as characteristic features of glaciation. The most important features developed by the erosional work of glaciers are as follows:

(a) Cirques or Corrie

These are a deep rounded hollows having steep sides, formed through erosion by snow and ice, thus are the chief characteristics of regions that have been glaciated. In regions where glaciers still exist, 'ne'va' often lies in the cirques which is reinforced by snow from the heights above and feeds the glacier. In regions where glaciation took place in the distant past, numerous lakes or 'term' fill the cirques and enhance the beauty of the mountain scenery.

For cirques the term 'Cum' is used in Wales and 'corrie' is used in Scotland.[51]

The cirques are circular and often open at one end. They are collecting basins for the nourishment of glaciers. Some geographers like Prof. Salisbury and Chamberlain do not take cirques as the erosional action of glaciers. They maintain that the cirque is scooped out as a niche by intensive frost action and subsequently enlarged by the same process. These hollows are filled with ice melt water (lake) and thus whole feature appears like a big armchair.

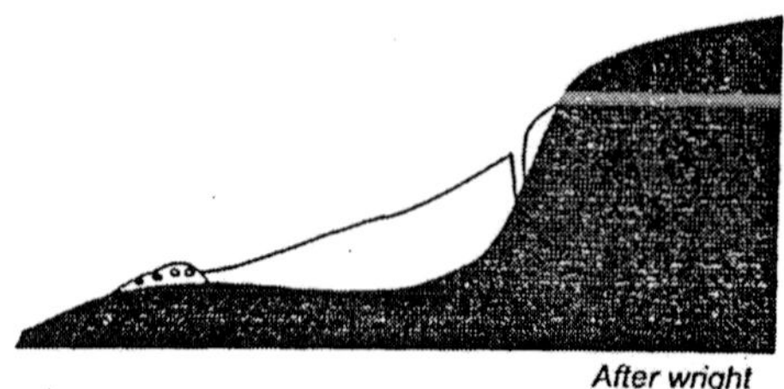

Fig. 4. A Corrie Glacier, with Berg-schrund and Moraine.

(b) Cols

It is a depression in a range of mountains or hills, generally providing a pass through the range. It is often produced by the continuing erosion of arete (a sharp mountain ridge) between two cirques on opposite sides of a ridge.

(c) Horns

It is a pyramidal peak that is formed when adjacent cirques have been cut back into a mountain and the faces being bounded by sharp and steep mountain ridges. In other words, number of cirques are sometimes found round a common ridge and they go on eating away the intervening wall. Thus such a feature is developed which appear like a tooth known as horn. A classic example of horn is Matterhorn in Switzerland.

(d) Comb-ridges

As the cirques grow, develop and eat back their backwalls, they produce a grooved upland with furrows. After some period these are reduced to a floor with terraces with steep walls and sharp edgeded ridges (aretes), also known as 'grats'. When these rock pinnacles (aretes) are horizontally arranged, they are called as combridges.

(e) U-Shaped Valley

The glacier cannot dig a new valley like running water. It can only widen the V-Shaped valley, converting it into the broad flat-floored trough or U-Shaped valley with steep walls. The glacier tends to iron out the irregularities of the side walls. Hence the valley profile becomes U-Shaped. St. Lawrence valley of America is a classic example of it.

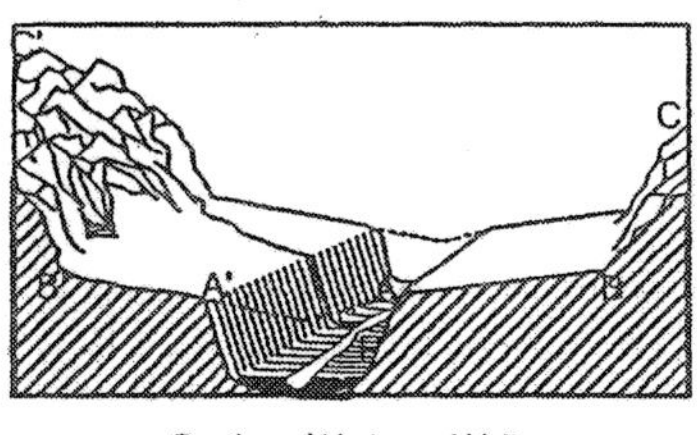

Section of U-shaped Valley

Fig. 5. Dotted line A,A' shows pre-glacial valley before it was over-deepend by the passage of a glacier. BC, B'C are the mountain slopes which rose above the glacier; AB, A'B' the shelves now covered with glacial debris forming pastures—the *alps*. Streams fall over A and A' into the main valley. P is the probable site of a power-station.

(f) Over-Deepened Valley

The over-deepend valley is formed in due course of time by the longer stay of glacier in the valley. Because the longer is the stay of the glacier, greater would be the widening and deepening of the trough, is the natural rule of glaciation. These features are quite common in Alpine valley which are known as 'alps' in Switzerland.

(g) Hanging Valley

These valleys are formed due to differential nature of glacial erosion. The main glacier with a greater ice supply is able to erode deeper than the smaller tributary glacier. Hanging valleys are common in a glacial regions, when a glacier has eroded the main river valley, the opening of a tributary into the later is left at some height above the new, lower valley. It thus forms rapids and waterfalls down the slope.

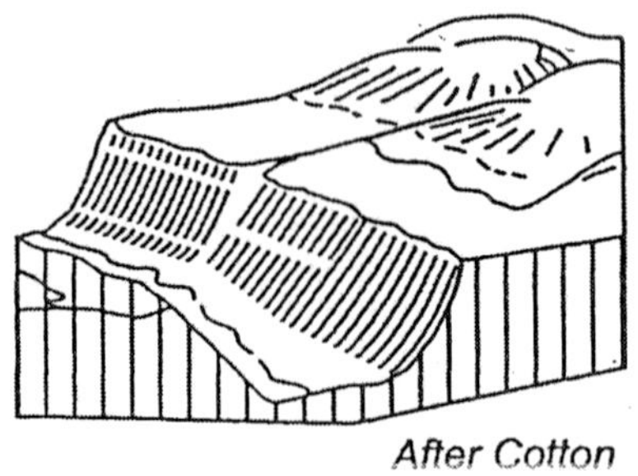

After Cotton

Fig. 6. A Hanging Valley.

(h) Rock Bastion

When the tributary glacier cut the steep walls into a series of steps which are called rock bastions. When the wall is exceptionally steep, these are known as rock bars. Alpine valleys are known for rock bastion. These steps are as high as 300 to 500 feet and as one goes up the valley the height goes on increasing.

(i) Roche Mountonnee

This is a resistant residual rock hummock, the surface of which is striated by ice movement. Its upstream side is smoothened by abrasion and its downstream is rough and much steeper due to plucking. The irregularities of the rock floor are smoothened and rounded into swinging curves and dome-shaped bosses known as Roche moutonnee or sheep rocks. The feature resembles a sheepskin-wig once worn in France, found in both highland and lowland glaciated region.

(j) Crag and Tail

The crag is a mass of hard rock with a precipitous slope on the upstream side, which protects the softer leeward slope from being completely worn down by the on-coming ice. Therefore, it has a gentle tail, strewn with the eroded rock debris. The Castle Rock of Indinberg (Scotland) is the classic example of this glacial feature.

DEPOSITIONAL LANDFORMS

The rock material carried by the glacier is deposited as soon as the ice begins to melt and the glacier terminates. The matter deposited at the end of the valley glacier is in the form of a ridge and is called moraine. These are the most widespread features of glacial deposition and are characteristic of both Alpine and continental glaciers. Moraines may be several hundred feet high and many miles long.[52]

The moraines consist of intermingled material of heterogeneous shape and size. They consist of fine rock floor and also of big boulders (40 to 50 feet dia) of several tons in weight. The material deposited is often loose and broken having no connection with the basement rock of the region where they lie. In one ridge there may be as many as twenty or more different types of rocks with different colours and textures. Lack of homogeneity is the chief character of moraines. There are various types of moraines according to the place where they have been laid or deposited. Some of them are as follows:

(1) Ground Moraines

The moraines which are deposited upon the ground or the bed rock upon which a glacier is moving are called ground moraines. They are thin and their surface is irregular due to uneven deposition of rock materials carried by the glacier because whatever load of rock debris the glacier cannot carry, is deposited on bedrock over which the glacier flows and this results in uneven deposition.

(2) Lateral Moraines

These are the moraines, which are deposited along the margins of the glaciated valleys and are ridge like accumulations of rock debris which are sometimes 30 m. (100 ft.) or more in height. They often persist for sometime even after the disappearance of glacier.

(3) Medial Moraines

These are ridge - like accumulations of rock debris formed along the middle part of a glacier. These originate by the union of two lateral moraines of two intersecting valleys. They are transitory in existence and are easily removed away because of their central position.

(4) Terminal Morains

These are accumulations of rock debris at the end or terminus of a glacier. They are of small width but very high sometimes 30 meters. Their length is determined by the width of the valley floor. They present a convex side downward which is due to faster movement along the central part of a glacier.

(5) Ablation Moraines

Ablation moraines are those fragments of rock that are being transported on the surface of the glacier. These are trains of rock material either parallel to the direction of ice flow or parallel to the margins. These consist of rounded pebbles and rock debris which were formerly embedded in the bottom layers but have since come to the surface. These fragments come to be deposited either when they slide down from the front of the ice or when the margin of the ice flow oscillates.

(6) Recessional Moraines

These are similar to terminal or end moraines except in their position. These are those which are left behind at the different stages of the retreat of a glacier. At every stage of retreat there are periods when the glacier halts and form a recessional moraines. In coarse of time these are deformed by the agents of weathering.

(7) Drumlins

In glaciated region, the ground moraines consist of low mounds of clay which sometime contains cores of bed rock. These small hillocks are of the shape of inverted teaspoons or half an egg, called as drumlins. They have their long axis parallel to the direction of ice movement. They are about 100 miles long and 200 feet in height.

(8) Till

It is a more or less uniform sheet of the debris brought by the glacier when its front retreats steadily. These occupy thousands of sq. mts. in the middle west of the USA. The depth being hundreds of feet of beds of sand and gravels. It is also known as glacial drift in Europe.

FLUVIO-GLACIAL DEPOSITION

These features show the combined effects of glaciers and ice melted water. The following are some of the chief features.

(a) Outwash Plain

The glacial streams carry huge quantity of rock debris, which form fan-like plains beyond the terminus of glaciers. They are also named as valley trains or overwash plain. The coarser sediments are deposited near the terminus while the finer sediments are carried farther downhill.

(b) Kames or Kame Terraces

Regions covered with glacial drift often present more or less flat topped, steep-sided and irregular elevations called kame terraces. These hillocks are built of steam-borne deposits, hence show rude stratification. They originate usually between the valley walls and the tongues of glaciers by sedimentation.

(c) Eskers

These are winding steep-sided ridge like features built of steam borne drift. The word 'esker' has been derived from Irish word meaning a path. Their origin may be due to the sedimentation in a subglacial tunnel.

(d) Crevasse Fillings

These are short ridges generally straight and are formed by filling up of crevasses by stream borne rock debris.

(e) Varves

These are layered clays alternating with coarser and finer sediments. In glacial regions the lakes are filled with summer and winter coarser and finer deposits layer after layer every year so long as the lakes persist.

CAUSES OF GLACIAL ACTION

The climate of a place is chiefly controlled by the solar radiation but it is also influenced by latitude, altitude, ocean currents, winds, rainfall and distance from sea. Glacial climates are therefore explainable by a less amount of solar heat. It may be more clear from the following explanations:

(1) Elevation

It is an observed facts that at higher altitudes the temperature is low. A fall of 1 degree C is noticed for every 450 ft. of altitude. It is because of less and dust - free (dust particles) air in upper layers.

(2) Carbon Dioxide

The greater the amount of carbon dioxide in the atmosphere, the more is the increase in temperature.

(3) Volcanic dust

A cloud of volcanic dust will prevent the solar heat to reach the earth's surface by reflecting the solar heat, thus cause a lowering of temperature.

(4) Moisture

The more the amount of moisture in the atmosphere, the more is the increase of temperature. The moisture laden clouds prevent radiation of heat room of the surface of the earth, thus causing an increase in temperature.

HUMAN ASPECTS OF GLACIATED LANDFORMS

Effects of glaciation on both landforms and human activities have profound influence in many parts of the world. In temperate regions of Europe and N. America they have their striking impact. The glaciers are still shaping the landscape in high mountain regions of the world in Alps, Andes, Rockies, and Himalayas. Somewhere their effects are favourable and sometimes unfavourable, depending on the intensity of glaciation, the relief of the region and erosional and depositional nature.

In hilly regions such as the mountain slopes of Scandinavia, ice sheet and glacier have removed most of the top soil, leaving them quite bare of vegetation. Soils are not capable of supporting agriculture, as they are very thin. The benches or alps and glacial drifts of valleys are good grazing fields during summer. Cattle are driven up in summer in these pastures and in winter return to the valley bottom. This animal migration is known as 'transhumance'. The extensive plains of East Anglia and central USA with a good deposition of humus are favourable for better farming. On the other hand there is an infertile sandy and gravelly outwash plains of north Germany and marshy boulder clay deposits of central Ireland and barren ice-scoured surfaces of Canadian and Baltic Shields. The presence of numerous erratic and perched blocks in parts of Britain, Alberta and Canada are the great hinderence in the way of farming and use of machines. But the old glacial lake beds with their rich alluvium deposits support rich cropping. The great lakes of N. America serve as excellent water ways. The overflow channels of them make natural routeways e.g. the Hudson-Mohawk Gap which links with Atlantic seaboard of USA. While the dominant drumlins region have poor drainage.

The fluvio-glacial deposits are economically very significant. Esker and Kames of outwash plain provide sand and gravel for building construction and highways. The glaciated lake-basins provide natural reservoirs. The waterfalls, hanging valleys and glaciated uplands are harnessed to provide hydroelectric power in Canada, Switzerland and Scandinavia. As a matter of fact, it has boosted the industrial growth in the region, particularly chemical and metallurgical industries. Besides this, glaciated mountains provide magnificent scenery for tourists attraction. Skiing, mountain climbing and sight-seeing are quite popular in Alpine e.g. France, Italy, Swiss Alps. On the other hand, recessional and terminal moraines are of no use to man as they comprise coarse material.

10
Karst Topography

UINDERGROUND WATER

The rain water that falls upon the surface of the earth, is naturally divided into three parts. One part is quickly evaporated, another part flows over the surface of the earth and the third goes underground. The part which flows as streams on the earth surface is called the 'immediate run off'. The water which is soaked by the earth is stored underground as subsurface water plays an important role in the formation of topography. A part of this water again comes upon the surface of the earth through springs and wells, which constitutes the 'delayed run off'.

UNDERGROUND CIRCULATION

At a certain depth below the surface all the rocks are saturated with water which means that the rock spaces are completely filled up with water. The upper surface of this saturated area is called 'water table'. The area below water table is called the 'zone of saturation'.

The percolation of water is chiefly controlled by the forces of gravity, adhesion and capillarity. The gravity causes a downward flow of water. In pervious beds like those of sand and gravel, the movement is rapid while in impervious beds of shales the flow is standstill. The molecular forces of adhesion tend to retain the water in the interspaces of the rocks. By the capillary action some water is sucked up through the minute holes of the rocks underground.

WELLS AND SPRINGS

The supply of water can be obtained from the wells which are dug sufficiently deep to reach below the water table. The sedimentary rocks can yield water very easily in comparison to others.

Artesian wells are the wells of special type. The word artesian has been derived from the name of a place 'Artois' in France, as the well of first of its kind was made here in 12th century AD. The permeable layer like that of sandstone overlain and underlain by impermeable layers like those of shale are all bent concave upward. At the outcrop of the sandstone layer water enters into it, but it cannot go out because of the presence of two impermeable rock layers on both sides of the sandstone. Water, therefore, collects in the curved layer of sandstone. This accumulation develops a hydrostatic pressure on the water, as the water keeps its level. Thus if a hole is made or a fissure permits the water to come out of the depression of the curve, it flows out with a great pressure, called as artesian well. The depth of these wells varies considerably. Some of them are just few meters deep, whereas others are hundreds of meters deep. Many small towns and villages depend on them for their supply of water. They are valuable for irrigation in semi-arid regions.

Springs are natural openings through which water flows to the surface. It is a continuous or intermittent flow of water from the ground. It is produced when rain water sinks through the ground to an inclined bed of impermeable rock, where it accumulates and finally gushes out through some crack or fissure. The accumulated water is led by fissure to the surface at some distant point below the level of intake. The water of springs may be cold or hot, soft or hard. The water becomes hot due to the chemical reaction or by touching hot 'magma' in the interior of the earth. The underground water acquires a warm or cold nature due to the warming and cooling of the surface of the earth in summer and winters respectively.

Spring water contains some dissolved matter, more common being bi-carbonates, chlorides and sulphates of calcium and magnesium. Some gasses, radium, iron, aluminum and potassium salts are also at times present. The spring water has some specific characteristics due to these dissolved substances. The soluble bicarbonates, chlorides and sulphates of calcium and magnesium make the water hard. The sodium and other salts impart some medicinal properties to the spring water, thus used as remedy for skin diseases, gout and rheumatism etc. In some springs water is hot and at times reaches boiling point. The author has an experience of boiling rice in it at Renie (Marwah) in J&K State. When a hot spring ejects water and steam in a column of sufficient height, it is known as 'geyser'.

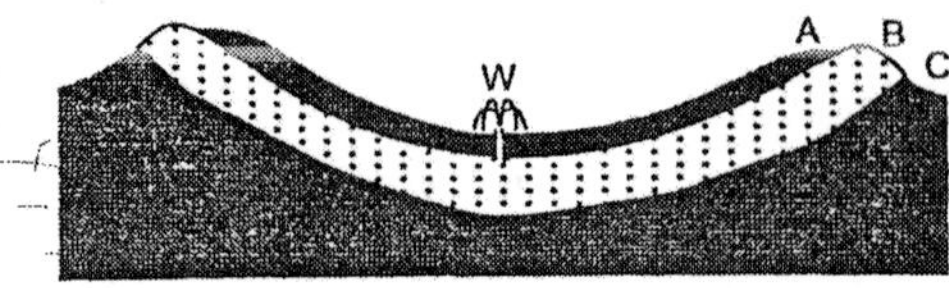

Fig. 1. A Type of Artesian Well: B is a permeable layer, A and C impermeable layers, W is the well.

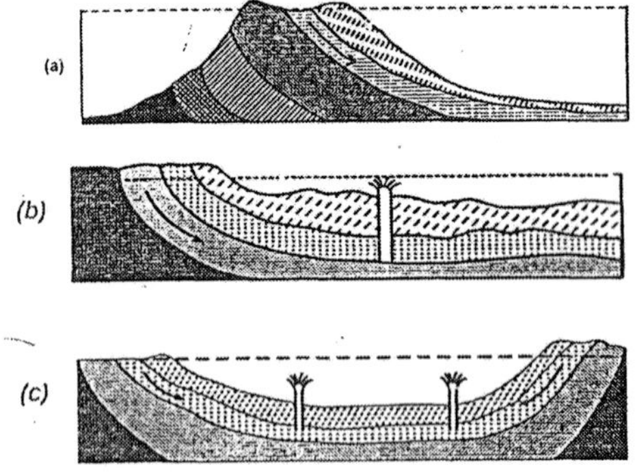

Fig. 2. Underground water and Artesian well—(a) a non-flowing well (b) well, in which water has to be taken out by pump (c) A self-flowing well—Artesian well.

UNDERGROUND WATER EROSION AND RELATED FEATURES

The mechanical erosion is almost impossible by underground water due to its very slow movement. When the soil is highly inclined, becomes saturated, the water acts as lubricating agent for it. Thus facilitating the flow of overlying rock mass, called as rockslides, which is disastrous in mountain region. Their deposits at foot-hills are called 'scree' or 'talus'.

The chemical erosion of underground water is highly important. The solution effects of it are conspicuous in limestone regions. The following important structures are formed as a result of underground solution of rock.

(a) Lapies

The surface of limestone rocks in a 'Karst region' that are grooved and fluted due to the solution by rain water that contains carbon dioxide are known as lapies in French. The grooves and flutings being the channels through which the latter runs down the rock faces. The ridges and furrows are known 'clints and grikes' in England, 'Kerrens' in Germany and 'bogax' in USSR.

(b) Sinks

These are saucer-shaped depression in the earth's surface in limestone region through which the surface stream may disappear and begin to flow beneath, along an underground course. These are the widened form of lapies, known as sinkholes, swallow holes or pot-holes.

(c) Dolines

The sinkholes allow the surface waters to flow downwards so that the swallow holes become very wide giving birth to dolines. These are funnel shaped hollows of varying sizes, which are made in limestone regions by the underground solution. When surface water mixed with CO_2 percolates downward through chalk region, a good amount of rock is taken into solution. The consequent collapsing of the roof and formation of basin is the result of this solution. Hence, naturally the sinks take the form of funnels but other shapes are also formed depending upon the structural peculiarities of the rocks.

(d) Caverns

Small caves and underground channels are sometimes formed

in limestone regions by the solution of underground rocks with the collapsing of the roofs the streams flow through these caverns and channels, which are sometimes as long as 30 miles underground. The caverns are of two types (1) horizontal chambers or galleries and (2) vertical shafts which are known as ponors.

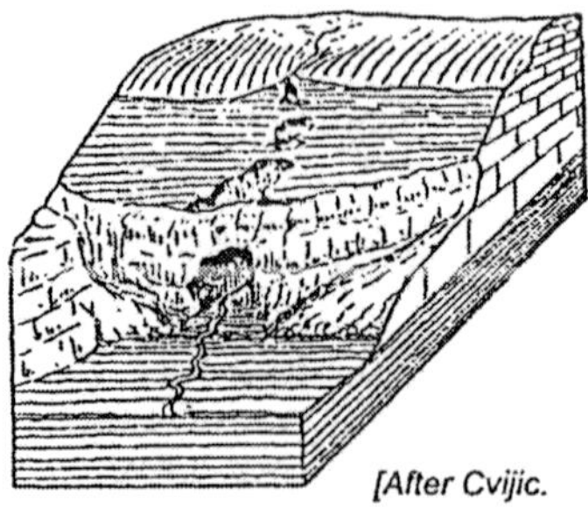

[After Cvijic.

Fig. 3. Incipient Stage of Valley Formation by Collapse of Caverns.

(e) Poljes and Hums

Due to the collapses of the roofs of the caverns huge depressions are formed. It is a flat bottomed huge depression, several kilometers in length formed mostly by solution of limestone in karst region. In the midest of these vast depressions residual masses of limestone rise here and there known as 'hums'.

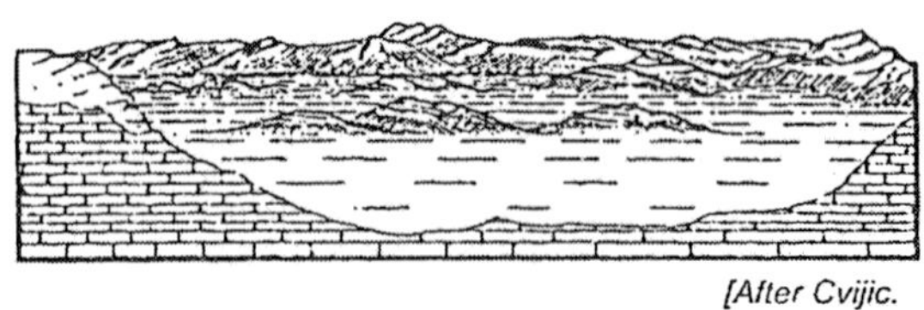

[After Cvijic.

Fig. 4. A Polje with Residual Limestone Masses.

(f) Natural Bridges

In a limestone country, it is a remnant of the roof of an underground cave or tunnel. Sometimes the roofs of these caverns are incaved and a bridge or arch of natural, tough and unjointed rock is formed known as natural bridge. These are semi-circular in outline. Virginia limestone arch, USA 66m. high with a span of 27m. carrying a road is a typical example of the natural bridge.

TRANSPORTATION

The dissolved material is carried in solution underground, till it is deposited anywhere. Sometimes they are carried as far as in seas and oceans or lakes through underground streams. It increases the salinity and lime to the ocean and lakes. The erosional and transportational work goes on side by side in underground caverns.

DEPOSITION AND RELATED FEATURES

The mineral matter hanging from an elevated point, which resembles an icicle in appearance is a limestone pillar. These are calcareous deposits formed by the percolating water containing calcium carbonate in solution from the roof of a limestone cave. As a drop of this solution hangs, the water partly evaporates and leaves behind a small quantity of calcium carbonate. Then the next drop leaves a further small quantity and the deposit grows downwards from the roof till it begins to hang like a pointed pillar. This is known as 'stalactite'. It looks like concentric rings of limestone placed over one another, tapering downwards.

'Stalagmite' is a column of calcium carbonate produced on the floor of a cave by water that contains calcium carbonate in solution falling from the roof. Some of the water evaporates, leaving behind the lime stone which grows as an upward pillar. It is generally shorter and thicker as compared with stalactite, but at its foot it has often a crater - like depression. Occasionally both these limestone pillars meet, producing a complete pillar from floor to roof of the cave. They attract a large number of visitors from far and wide for their beauty and design particularly in England and Australia.

KARST CYCLE OF DEVELOPMENT

The name karst for an area of rugged limestone plateaus and ridges near Adriatic coast of Yugoslavia was first applied by J.W. Beedi in 1911. The concept of cycle of erosion was also applied by Jovian Cviji in 1918 and W.M. Davis in 1930. Karst cycle of erosion becomes more operative over two types of structures (a) folded limestones and (b) faulted beds of limestones. It becomes more effective where thick beds of limestone are exposed on the ground, where rainwater directly dissolves it. The characteristic features of different stages of karst cycle of erosion are given as follows:

(1) Youth

Karst erosion starts with surface drainage in limestone region. The rainwater mixed with carbon dioxide becomes an active agent of dissolution. It reacts with limestone along the joints, sinks and swallow holes. These holes gradually increase in number and size due to continuous solution giving rise to dolines. The surface streams disappear through these dolines and sink holes. The ground surface is characterized by rough terrain due to the development of 'lapies', because of dissolution of limestone along their joints. The underground drainage starts the formation of caves and caverns.

(2) Maturity

The increased solution of carbonate rocks results in gradual enlargement of caves, galleries and passages. The covering roofs of caverns undergo the process of thinning which causes their collapse giving rise to natural bridges, uvulas, poljes and karst windows. The residual uplands with highly pitted surface between uvulas become ridges. The late maturity is characterised by the destruction of most of solutional landforms.

(3) Old Age

In the last stage the roofs of the caverns totally collapse which results in the formation of a huge depression. This depression is known as polje. The residuals of carbonate rocks projects slightly above the ground called hums. The subterranean drainage again appears on the ground surface of the earth. All the features produced during the first and second stage are eliminated.[53]

THE MAJOR LIMESTONE REGION

The most characteric stretch of limestone occurs in N.W. Yugoslavia. Other regions include the Causes district of southern France, the Pennines of Britain, Yorkshire and Derbyshire. Besides

Kentucky region of the United States, the Yucatan Peninsula of Mexico, the Cockpit country of Jamaica and the limestone hills of Perlis.

HUMAN ACTIVITIES OF KARST REGIONS

Karst regions are often barren and at best carry a thin layer of soil. The vegetative growth is very poor due to the porosity of the rocks and absence of drainage. Thus short turf and very poor grasses are the characteristic features of limestone region. It supports some sheep grazing in the tropical regions of heavy rainfall all the year round. The population is very sparse and the settlement are often scattered in limestone region. The important mineral of the region is lead. The limestone is also used as building material or quarried for the cement industry.

The landforms of chalk are rather different from those of other limestones. There is little or no surface drainage and valleys which once contained rivers are now dry. These are often called 'combers'. The chalk forms low rounded hills in south-eastern England and northern France where they are called domes. The chalk is covered with short turf and in places of heavy rainfalls with woodland. The region is often used as pasture land and a little arable farming is also possible.

11
Coastal Landscape

COASTAL LANDFORMS

Coastal landforms are shaped in different features by the marine agents of erosion, transportation and deposition. The geomorphic changes brought about by these agents are apparent in a relatively very short period of a few decades. Thus ocean is an active agent of denudation. The term sea and ocean are synonymous and include the vast bodies of water existing between the continental lands.

MECHANICAL WORK OF OCEANS

The geomorphic work of the oceans is mostly mechanical and only a insignificant part is chemical. Mechanical work is two-fold, viz. erosive and depositional. The eroded material by sea waves on the shore and debris fallen from the coasts is deposited by the oceans. In the sea it is the body of moving water that perform erosion This is accomplished by waves, currents and tides.

The coastline under the constant action of the waves, tides and currents in undergoing changes from day to day. The waves do little damage to the shoreline on calm days (without wind). It may instead help to build-up beaches and other depositional features. In the severe storms the waves attain a depth of about 600 feet and reach to their greatest magnitude of erosive power. It is all at the shore that the destructive work is greatest. Movements of such intensity will wear down not only the cliffs but also sea walls. The tides and currents on contact with the shores, make very little direct attack on the coastline. The rise and fall of tides twice daily enlarges the zone of erosion by waves. On the other hand currents play an important part in shifting material away from the shore such as silt, sand and gravel.

THE MECHANISM OF MARINE EROSION

Waves are the most important and powerful agents of erosion. They are caused due to the blowing of winds over the water surface of the oceans. The water moves to and fro without covering any distance. The vertical height of a normal wave may be about 60 feet from 'crest' to 'trough', whereas its horizontal distance can be about 400 feet from crest to crest. During storms it is at its greatest magnitude depending on the speed and duration of the winds. Near the shores, in shallow waters the crests curl over and break into the shores in a mass of foam as breakers. The water that rushes up the beach and hurls rock debris against the land is termed as 'swash', and when it retreats it is known as 'backwash'. Marine agents of erosion operate in the following four ways:

(1) Corrosion

The corrosion is a process by which the base of the cliffs are weathered and eroded away by the waves of the oceans. The waves use the rock debris of different shape and size as its tools for its erosional work. The tides and currents share with waves by sweeping the eroded material into the sea.

(2) Attrition

Attrition is a mechanism by which the eroded material of the beach such as boulders, pebbles, shingle and sand are grind into very small pieces. The constantly moving waves hurl these fragments against one another till they are broken into fine sand which is the chief characteristic of beaches.

(3) Hydraulic Action

The waves splash against the coast, in their forward surge, may enter joints and crevices of the rocks. With this process the air imprisoned inside the joints is immediately compressed. The air expands with explosive violence when the waves retreat. The continuous such action enlarges the joints and the result is that the rock fragments are praised apart.

(4) Solvent Action

On limestone coasts, the solvent action of sea water on calcium carbonate sets up chemical changes in the rocks and disintegration takes place. This mechanism is limited to limestone coasts.

MARINE EROSIONAL FEATURES

The significant coastal topography is the result of marine erosion through its chief agents such as waves, tides and currents. The main features formed by solutional process and agents are caps, bays, cliffs wave-cut platforms, caves, arch, stack, stump goes and gloups etc. Some of them are treated in detail as under:

(1) Caps and Bays

The coastlines having the rocks of varying resistance are irregularly eroded by the continuous action of sea waves. This is particularly pronounced where hard rocks e.g. granites and limestones occur in alternate bands with softer rocks as sand and clay. The softer rocks are worn away by the waves to form a wide indentation into the land known as bays. On the other hand the harder rock persist as a prominent headland or promontory projecting into the sea called as cape. Persian Gulf and the Bay of Bengal are formed due to other causes such as submergence or earth movements.

(2) Cliffs and Wave-Cut Platform

A cliff is a high and extremely steep rock face that approaches vertical along the coast line. The rate of recession depends on its rock formation, stratification and jointing and their resistance to wave attack. If the beds dip landwards the cliff will be more resistant to wave erosion. On the other hand if the beds dip seawards, large blocks of rock will be dislodged and fall into the sea. Thus the cliff will rise in a series of steps.

As a cliff recedes landwards under the pounding of waves, an eroded base is left behind, known as 'wave cut platform'. It gently slopes towards the sea on which the abrasion remains continue until the debris and pebbles are swept away into the sea. The platform attains the width of miles together and is entirely covered with water.

(3) Cave, Arch Stack and Stump

The continuous and prolonged wave action on a cliff base results in the excavation of holes called 'caves'. When two opposite caves approach each other from either sides and unite, they give rise to an 'arch'. Sometimes the total collapse of these archs results by further erosion. The seaward portion of the headland will remain as a pillar of rock known as a 'stack', which may be hundreds of feet high. In the course of time these stacks or rock pillar are gradually removed by continuous wave erosion leaving behind only the mounds of these pillars, known as 'stumps'.

(4) Geos and Gloups

The occasional splashing of the waves against the roof of a cave may enlarge the cracks and joints by compressed air which is trapped inside. With the continuous process a vertical shaft is formed which may eventually pierce through to the surface. The wave forces the water and air through these shafts known as blow-holes or 'gloups'. When the roofs of these caves collapses, a long, narrow inlet or creek is developed. Such long and deep clefts (100 x100 feet) are called 'geos'.

TRANSPORTATIONAL WORK

The marine agents waves erode the material from the coastal region and tides and currents generally transport the same into the sea. The material is eroded and transported by the waves also from coastland towards the sea and from sea it is again transported towards the coast by the mechanism of backwash and breakers. The material involved in the transportation by sea waves include sands, silt, gravels, pebbles, cobbles and boulders.

MARINE DEPOSITIONAL FEATURES

(1) Beaches

The strip of land or terrace bordering the sea that slope gently downwards from the land. It is recognized as that part which lies between high and low water marks and produced by the action of the sea. It consists of shingle, sand and mud. A beach is the most dominant form of the constructive work of the sea.

The eroded material is transported along the shore in several distinct ways. The 'long-shore drift' which come obliquely to the coast carries the material along the shore in the direction of the dominant wind. The 'backwash' removes part of the material seawards along its bed and deposits it on the off-shore terraces. The finest particles of silt and mud are deposited in the shallow waters. The length of the beaches varies according to the configuration of the bordering land. They go for miles in plain areas where as in upland regions they are very short in extent, where the land descends abruptly into the sea.

(2) Spits and Bars

The long shore drift transport the material continually and deposit the same across the mouth of the rivers and bays. By prolonged process the material is added and piles up into ridge forming a tongue or spit. The one end of the spit is attached with land and the other projecting into the sea.

A ridge of sand and shingle that is formed in sea across the mouth of a river or the entrance to a bay or harbour is known as 'bar'. It lies approximately parallel to the coast, also known as off-shore bar or barrier beach.

(3) Marine Dunes

Dune is a low ridge or hillock of drifted sand mostly moved by the wind. With the force of the on-shore winds, a large amount of coastal sand is driven landward forming extensive marine dunes that stretch into due-belts. These dunes are as high as 130 feet, while the width is in thousands of miles.

COASTAL CYCLE OF EROSION

(a) Initial Stage

If a coast of moderate relief is submerged, the initial shoreline will be irregular, due to drowned river and glacial valleys. It forms fiords, separates headlands and promotories. It will be rougher and more irregular in detail.

But in case of shore-lines of emergence, they are straight and regular. Coastal plains extend for greater distance in sea water, with gentle gradient the breakers form 'notch' and 'nips'.

(b) Youth Stage

Continuous and prolonged erosion accentuates the relief. In early youth, prominent cliffs, erosion remnants, such as stacks, pulpit rocks, bridges are prominent. Further wear off of promotories and headlands result in the formation of high sea cliffs. The bays and mouth of rivers are filled up by sediments. Further erosion causes recession of headlands and lagoons.

(c) Maturity

Headlands are cut even further back than original embayments. Sea cliffs loose heights on account of weathering, which lowers general relief. Sediment laden bays are removed off. The shorelines begin to show irregularities are gentler than those of the infancy stage. The benches (wave cut) transform into wave built terraces, the height and gradient of coastland decreases significantly.

(d) Old Age Stage

In this last stage the land surface becomes low, the waves become weak, cliff lost their height, the least land derived sediments reach the shore. The shore aquires a gentle slope with prominent wave cut benches. The whole landmass is reduced to base level of marine erosion.

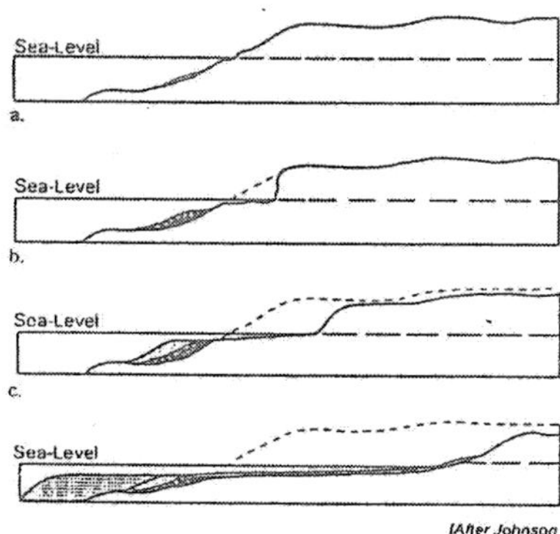

Fig. 1. Stages in the Development of a Shore-profile following Submergence. (a) Initial stage. (b) and (c) Stages of yough. (d) Mature stage.

12
Climatology

WEATHER AND CLIMATE

The science which treats of the various climates of the earth and their influence on the natural environment is known as climatology. The chief elements of atmosphere are atmospheric pressure, temperature, humidity, rainfall, cloudiness, precipitation, wind speed and direction. The condition of these elements at a certain time or over a certain short period is called the weather. The average weather conditions of a place or region throughout the seasons is known as climate. It is governed by latitude, altitude, position relative to continents and oceans, ocean currents, winds and insolation. Broadly speaking, the interiors and the eastern parts of the great continents have a continental climate, with small rainfall, low humidity and a great range of temperatures, both diurnal and seasonal. On the other hand oceanic islands and the western parts of continents have a heavier rainfall, higher humidity and more uniform temperatures. However, there are numerous exceptions as the local climates are still further modified by altitude, proximity of mountains etc.

The equatorial regions has a climate almost synonymous with weather as there is very little variation in the weather. The weather is often so variable that the climate can scarcely be described in a concise terms, between the tropics and poles, especially in the regions of Westerlies.[56]

In other words the climate of British Isles, is so changeable that many people have commented that "Britain has no climate but only weather". Conversely, the climate of Egypt is so static, that it makes a good deal of sense when people say that the 'Egypt has no weather, but only climate'.

DEFINATION AND SIGNIFICANCE

The climate has been derived from the word "Klima" meaning initially the curvature of the earth's surface. Greeks held the variations of climate to occur from north to south, which resulted in three climatic zones as torried, temperate and frigid. The scientific study of the climates or climatology is closely related to meteorology. It is dependent on meteorology for data regarding various elements of weather for effective explanation.[57]

The weather of a place is the condition of the atmosphere above it. It implies the conditions of temperature, pressure, wind, humidity, clouds and rain at a certain time. On the other hand, the climate is the average condition of weather at a place for a considerable period. The Chief difference between weather and climate is in the length of period of observation. Secondly, the weather is always changing while climate is a permanent and constant feature. Thirdly the weather denotes for actual conditions of the atmosphere where as the climate denotes for average conditions of the atmosphere.

The climate is very significant for mankinds as it has a profound influence over the activities of man in his everyday life. It influence the basic requirements of man i.e. food, clothing and shelter. The physical characteristics, the mental alertness and even our racial differences are closely related with climate. The farmers and their crops are still at the mercy of weather and climate despite the advancement of science and technology. The temperature conditions, humidity and precipitation directly controls the health conditions of both men and crops. These factors promote or discourage the growth of fungus and diseases. The death rates are normally high in tropical countries and low in deserts. The cool and fresh air of mountainous areas is always ideal for good health.

Though men are still unable to tame the forces of nature such as floods, droughts, typhoons or hurricanes, a sound knowledge of the trends or the weather systems can often help to avoid or reduce the seriousness of calamities. Thus meterorological stations scattered all over the globe using some of the most up-to-date weather instruments prepare weather maps and charts which are helpful for general awareness of the masses. This is why today the farmers are becoming more and more dependent upon this department to plan their work for the season. It assist farmers to take due precautions against frost, hail heavy snowfall or a period of possible droughts. Hence a fair knowledge of the weather is not only useful but often essential.

CLIMATIC ELEMENTS

The climate of a place is always dependent upon the combination of certain factors, which are known as climatic elements. The principal climatic elements which control and determine the climate of a place are as follows:

(1) Latitude

The angular distance of a point on the earth's surface north or south of the equator, as measured from the centre of the earth is called as latitude. The rays of the sun are vertical at the equator all the year round but they become slanting towards the poles. The result is that higher the latitude, less intense is the heat received from the sun. The slanting rays of the sun spread over a greater area of the surface hence they have less heat. Conversely the vertical rays cover a smaller area hence they have greater intensity in heat.

(2) Altitude

The vertical distance above mean sea level, usually measured in feet or meters is called an altitude. The higher the altitude of a place, the lower is the temperature and vice versa is a rule. This is because the air in lower elevation is often thicker while it becomes thinner as we go up. Thus, thicker the air, more is the capacity of insolation and thinner the air less is the capacity of insolation. The altitude controls temperature and rainfall.

The mountain chain give rise to the cold and dry winds. They also act as barriers to the rigorous climatic conditions and also prevent these winds from reaching the other side. The classic example of such mountain chain is the Himalayas which protects India from the cold waves of northern Asia. The intensity of solar heat also depends on the degree of slope, as more the slope less will be the heat. Thus, the temperature of a place decreases or increases according as the slope is towards the sun or away from the sun.

(3) Distribution of Land and Water

In winter the land is comparatively cooler than water while in summer the land is hot and water is cooler. This is due to the fact that water is heated and cooled more slowly than land. This variation and inequality of temperature has a significant influence on the weather and climate of a place. That is why the places in proximity to the sea have an equable type of climate all the year round, whereas the places far away from the sea have a continental climate. The continental climate is marked by extremes i.e.. the range of temperature is greater than the equable climate. The great difference of daily temperature causes land and sea breezes while the seasonal changes results in monsoons.

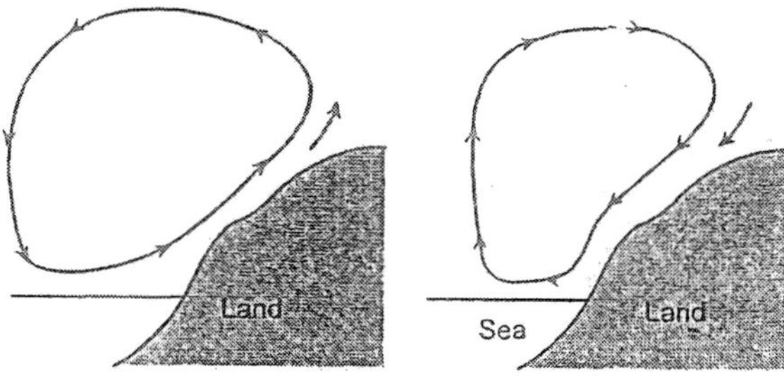

Fig. 1. See breeze-day. Land breeze-night.

(4) Ocean Currents

Currents are the gushing streams of water over and under the surface of the calm ocean water. The ocean currents have a very great influence on the climate of a region which they wash or approach. Usually the western margins of the continents are washed by warm currents moving towards the equator. These currents influence the climate of the adjacent lands. The Labrador current freezes the mouth of the St. Lawrence river which results in fall of temperature and precipitation in the surrounding areas. The Gulf stream, a warm current influences the climate of western Europe and British Isles while the Kuro Siwo does the same to Japan and west coast of N. America.

(5) Surface Covering of the Land

The surface of the earth is often covered with either vegetation or snow or it is bare. Each has its own capacity of insolation or radiation so accordingly the temperature of an area is modified. The bare soil is heated soon than the land covered with vegetation or snow. Thus the former land is warmer and the later cold.

(6) Prevailing Winds

The winds, indicated by direction, at a certain place or in a certain area which has considerably higher frequency than any other are known as prevailing winds. The winds blowing over a country exert an important influence on the climate of the place. The winds blowing from the sea bring moisture, conversely the winds blowing from land are dry. Similarly equatorial winds are warm while the polar winds are cold. The cyclones and anticyclones have a great influence on climate in temperate regions.

ATMOSPHERE

The surface of the earth is surrounded on all sides by an envelope of air and this gaseous sphere hanging over the surface of the earth is called as atmosphere. This blanket of air consists principally of a mixture of gases mainly oxygen, nitrogen and carbon dioxide and variable quantity of water vapours. The nitrogen, a chemically inactive gas, serves merely to dilute the more important oxygen. With increasing height above the surface of the earth, the atmosphere becomes more and more rare, but within the layers the relative proportions of the gases remain constant excluding water vapours. From the weather and climatic view point the relative humidity is of paramount significance besides the temperature and atmospheric pressure.

The atmosphere is of great significance to both human and plant life. We live on oxygen while the nitrogen is necessary for plant life. The temperature, pressure and moisture content of the atmosphere influence the physical environment of a place, which have an indirect influence on man. The solar energy passes through the atmosphere to reach the earth and its intensity of power is effected by this blanket of air.

Besides providing all necessary gases for the sustenance of all the forms of life on the earth, it also filters the incoming solar radiation. Thus prevents the ultraviolet rays of the sun to reach the earth's surface. These rays are poisonous and can burnt away all type of life on the earth if reach directly. The atmosphere is thousands of kms in height but the effective atmosphere is supposed only upto a height of 29 km. The lowest layer in which the weather is confined is known as the troposphere. It extends from the earth's surface for a height of 6 miles and within it the temperature falls with increasing altitude. All the phenomenon of climate exists within this layer. The air is thickest with the surface of earth and it thins out higher up.

The second layer above the troposphere is the stratosphere which extends upwards for another 50 miles. This layer is with extremely thin air with no weather elements, like clouds, dust particles, water vapours etc. Beyond this layer there is the ionosphere with several hundred miles up. The short radio waves are developed in this layer. Above this is the ozonesphere, a zone of azone gas, which prevents some of the rays to reach the surface which are injurious to health.

COMPOSITION OF THE ATMOSPHERE

The surface of the earth is covered on its all sides by a thick layer of air which is composed of three significant-elements viz. gases, water vapour and dust particles.

(a) Gases

Nitrogen and oxygen are the two principal gases constituting 99% of the atmosphere. The other remaining gases are only 1 percent. Nitrogen is chemically inactive and serves to dilute the oxygen in the atmosphere and checks destruction of plant life. Oxygen is a chief gas from the viewpoint of mankind and the animals, being essential to the process of breathing. Oxygen is also significant for combustion of burning matter. Carbon dioxide is used by green plants for photosynthesis. It absorbs most of the radiant energy of the earth and reradiates it back. Ozone absorbs most of the ultraviolet rays of the sun which are dangerous for all type of life on the earth.

The Gases Present in the Atmosphere

S No.	*Gasses*	*Percentage*
1.	Nitrogen	78
2.	Oxygen	21
3.	Argon	0.93
4.	Carbon dioxide	0.03
5.	Hydrogen	0.00005
6.	Neon	0.0018
7.	Helium	0.0005
8.	Ozone	0.00006
9.	Krypton	(Trace)
10.	Xenon	(Trace)
11.	Methane	(Trace)

(b) Water Vapours

The main sources of water vapours present in the atmosphere are the water bodies in the form of oceans, seas, lakes, rivers and vegetation cover of the earth. As it depends on temperature, it decreases from equator towards the poles. Water vapours in the gaseous form or minute particles of water suspended in the atmosphere acts as a blanket of heat while on condensation it gives rain, hail and snowfall. For various types of weather phenomena like fog, mist, clouds etc. in the atmosphere, the water vapours are responsible.

(c) Dust Particles

A solid matter consisting of very minute particles, occurring everywhere in the atmosphere are known as dust particles. They are so minute that they are not visible with nacked eye. Hygroscopic dust act as nuclei for the condensation of water vapour in the atmosphere, thus help in the formation of water drops, clouds and various types of precipitation. The dust particles, salt particles, pollen, volanic ashes, smoke and soot are together termed as particulate matter, which are kept in suspension by air. This matter also helps in the scattering of solar radiation adding charming colours of red and orange at sunrise and sun set. Due to the selective scattering of solar radiation by dust particles the sky appears blue in colour.

STRUCTURE OF THE ATMOSPHERE

On the basis of the modern knowledge the height of the atmosphere is estimated at between 16000 to 29000 km from the sea level but it is most significant only upto a height of 800 km. Upto an altitude of 5.6 km lies about 50 percent of the atmosphere while 97 percent of it is confined to a height of 29 km. On the basis of varying characteristics the atmosphere can be divided into four layers (spherical shells) or zones as follows.

(1) Troposphere

It is the lowest zone of the atmosphere extending upto a height of 16 km over the equator and about 6 km over the poles. Within the troposphere the air temperature normally falls with increasing height at the rate of 6.5 degree C per thousand meters or 0.6 degree C per hundred meters. This decrease of temperature is known as 'normal lapse rate'. Almost all the weather phenomena such as fog, cloud, dew, frost, hail, cloud storm and lightning occur in this very zone. Thus it is the most significant for all type of life on the surface of the earth. In the atmosphere, the boundary between the first and second layer is known as the 'tropause' or the ceiling of the troposphere. The atmospheric disturbances are no longer visible as one crosses the tropause.

(2) Stratosphere

It is the second layer of the atmosphere, extending upto 50 km on an average from tropause. Within the stratosphere, the temperature is almost constant in the vertical direction. This is why sometimes the term "isothermal layer" has been applied to it. However, it is misleading as the temperature changes in a horozontal direction and to some extent with height also in lowest layers of this zone. There is no disturbance of weather element except some cirrus clouds and a feable winds.

The upper portion of the stratosphere which has a maximum concentration of ozone gas is called ozonosphere. This layer extends upto 55 km, absorbs the extremely hot ultraviolet rays of the sun. Thus, it acts as a protective cover for the biological communities in the biosphere. This zone is filled with a faintly blue irritating gas with a characteristic pungent odour. Recent researches have shown that there is a gradual depletion of ozone gas in the atmosphere. The main culprits of ozone destruction are halogenated gases which belong to the synthetic chemical category like chlorine, fluorine and carbon. Chlorine reacts with water and thus, depletes ozone which can result in rise of temperature of ground surface. This can cause global warming, acid rain, melting of glaciers, rise in sea level, skin cancer etc.

(3) Mesosphere

This zone extends from 55 km to 80 km. In this layer the temperature again decreases with increasing height. In the upper most limit of mesosphere temperature becomes 80 degree C.

(4) Thermosphere

It is the fourth layer or zone beyond mesosphere in which the temperature increases with increasing height. According to an estimate the temperature of the upper limit of it becomes 1700 degree C. This zone is further sub-divided into two layers viz (a) ionosphee and (b) exosphere.

(a) Ionosphere

This layer extends from 80 to 640 km reflects the singals of low and high frequency radio waves, back to the earth. This layer is associated with solar radiation so disappears with the sunset. The ultraviolet rays from the sun splits up or ionizes atoms of the air, forming the ionosphere.

(b) Exosphere

It represents the upper most layer of the atmosphere. The temperature increases to an intensity of 5568 degree C in this layer. It extends beyond 640 km but the outer limit is not known as very little is known about this layer.

INSOLATION

The radiant energy received from the sun by the earth and other planets is called insolation. It varies considerably over different parts of the earth's surface. The amount of insolation reaching any place during any one day depends on solar constant. In other words it depends on the area of the surface and its inclination to the sun's rays, the transparency of the atmosphere and the position of the earth in its orbit. This amount varies throughout the day with the changes in the angle of incidence of the sun's rays and the length of the day. The insolation is, however, greatest at the equator, it decreases at first slowly, then more rapidly, then slowly again towards the poles. It shows the least variation throughout the year at the equator and varies considerably at the poles.[58]

The sun is continuously radiating heat energy into the space. But, because of its small size and great distance from the sun, the earth receives only a minute percentage (one in two billion parts) of it. Yet this small quantity is of great importance, as it controls all the physical and biological phenominon of the earth. The amount of insolation reaching the earth's surface in the form of short waves depends on the following factors.

(a) The angle of incidence or inclination of sun's rays.

(b) The duration of sun shine or length of days.

(c) Transparency of atmosphere.

The length of day light and angle of incidence together varies with season and with latitudes, which determines the amount of heat received by the earth surface. The amount of solar radiation also depends on the transparency of the atmosphere. The total annual insolation is maximum within the tropics and it gradually decreases towards the poles. Along 45 degree latitude it is only 75 percent of that of equator. It is reduced to 50 percent along Arctic and Antarctic circles and less than 40 percent at the poles.

The chief and only source of energy is the sun for the atmosphere of the earth. It has a surface temperature of more than 10800 degree F which travels through space for a distance of 93 million miles to reach the earth. It is estimated that of the total radiation of the sun 37 percent energy is lost by reflection back to space by dust, clouds and air molecules. Another 12 percent is absorbed directly by water vapours and gases. Thus the remaining 51 percent is able to reach the earth. According to an estimate every minute the earth is receiving 1.94 calories of heat per sq. centimeter. This solar constant differs from place to place depending on altitude of the sun, the amount of atmosphere to be crossed, duration of sun light and output of solar radiation.

The perpendicular rays have to pass through a smaller portion of the atmosphere and spread over a smaller extent of surface. Hence the regions receiving indirect and slanting rays are cooler. The classic example being the polar region than equator which receive direct rays and are warmer. The solar radiation warms the earth's surface and in turn, the earth warms the air above it by direct contact known as conduction. The air is also warmed through the transmission of heat by upward movement of air currents called convection.

TEMPERATURE

The degree of heat of a body usually expressed in degrees on the centigrade (C) or the fahrenheit (F) scale measured by a thermometer is called temperature. The temperature decrease in general from equator towards the poles. But the temperature of any particular place depends also upon other factors like altitude, proximity to the sea, prevailing winds etc.

HORIZONTAL DISTRIBUTION OF TEMPERATURE

The lower the latitudes, higher is the temperature and vice versa is the rule. Thus, according to it temperature decreases from equator towards the poles. This poleward decrease of temperature is called temperature gradient. But as a matter of fact the highest temperature is recorded in the tropics of cancer and capricorn. Because a sizeable portion of incoming solar radiation is reflected by clouds and lost by evaporation near the equator.

The surface distribution or horizontal distribution of temperature depends on isolation, land and sea, winds, currents and seasonal changes which is represented by isotherms. Isotherms are the lines drawn on the map joining places having the same temperature at a particular instant, or having the same average temperature over a certain period. The temperature is normally reduced to mean sea level in order to eliminate differences due to altitude. Isotherms are the lines joining places of same temperature, especially for January and July showing the average temperatures of mid winter and mid summer. These lines normally run east - west and are generally parallel to latitudes. These are straight but they bend at the junction of continents and oceans due to differential heating and cooling of land and water. Isothermal lines are more regular in southern hemisphere due to the dominance of oceans as compared with northern hemisphere. The closely spaced isotherms in northern hemisphere denote rapid rate of change of temperature and steep temperature gradient and vice versa. The isotherms passing from the land towards ocean bend equatorward during summer and poleward during winter. The January and July isotherms are taken to represent horizontal distribution of temperature during winter and summer respectively because they represent seasonal extremes.

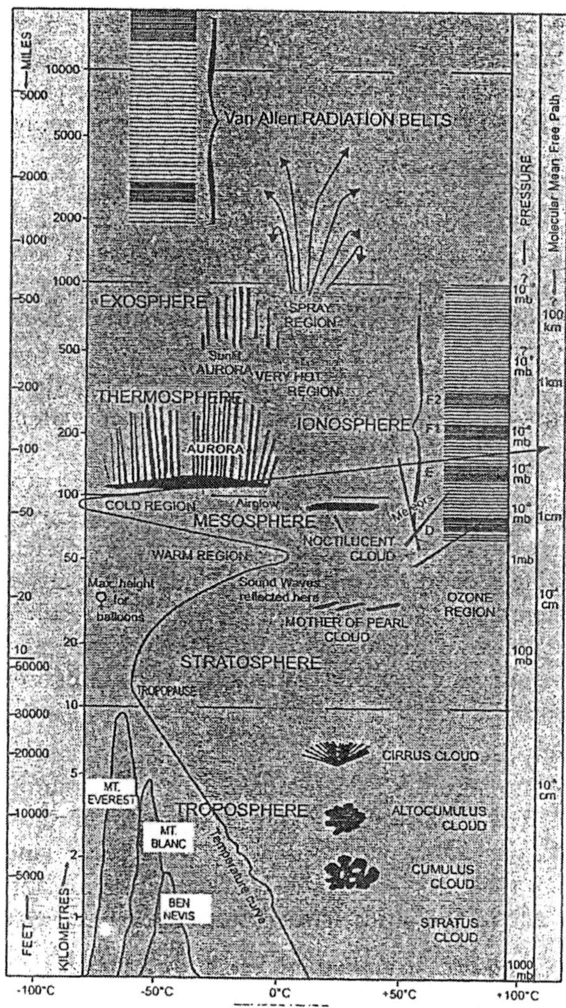

Fig. 2. Temperature

On the basis of surface or horizontal distribution of temperature the whole earth has been divided into three zones which are as under:

(1) Tropical Zone

The region lying between the tropic of cancer and tropic of capricorn 23.5 degree north and south from equator sometimes known as torrid zone. The sun's rays are more or less vertical on the equator throughout the year. The other areas also get direct rays for at least once every year. Thus around the equator there is no winter because of high temperature all the year round.

(2) Temperate Zone

This zone extends between the tropic of cancer and Arctic circle and tropic of capricorn and Antarctic circle in northern and southern hemisphere respectively. In other words temperate zone lies between 23.5 degree and 66.5 degree in both the hemispheres. There are marked seasonal contrasts with the northward and southward migration of overhead sun. Thus, the range of temperature between summer and winter become extremely high.

(3) Frigid Zone

This zone extends between 66.5 degree latitude and 90 degree latitude in both the hemispheres. Frigid zone is characterized by more slanting or oblique rays of the sun. Thus the regions experience very low temperatures throughout the year. The result is that the ground is covered with snow and ice due to the temperature mostly below freezing point. The duration of days and nights are 6 months each on both the poles.

It may be pointed out that the Greeks gave the un-due importance to latitudes in determining the temperature zones as described above. They over looked the other significant factors of control like prevailing winds, ocean currents, relief, land and sea etc. Soupan took all these factors into consideration and divided the globe into temperature zones on the basis of isotherms.

VERTICAL DISTRIBUTION OF TEMPERATURE

As a matter of fact the atmospheric temperature decreases with increasing elevation. But the rate of decrease varies with time, season and space. On an average the decrease of temperature with increasing altitudes is 6.5 degree C per thousand meters. This decrease of temperature is known as normal lapse rate or vertical temperature gradient. It is one thousand times greater than the horizontal lapse rate. The atmosphere is heated by the earth through the process of convection, compression (conduction) and radiation. Convection is a process of transmission of heat from one part of a liquid or gas to another by the movement of the particles themselves. When the lower portion of a mass of air is heated, it expands, its density is reduced and rises up carrying its heat with it to be replaced by cool air which in its turn is heated. In other words convection is the upward movement of air which has been heated by contact with the earth's surface. This rise of air is called convection current.

The air pressure is higher in the lower portion of the atmosphere near the surface of the earth because of the weight of all the air layers lying above. Thus, the density of air is maximum in the lower portions of the atmosphere which decreases rapidly upwards. The density or thickness of air or compressed layer of the lower atmosphere get heated up due to pressure from above and their temperature rises.

The third method of heating the atmosphere is that of radiation. The quantity of water vapours, dust particles, and carbon dioxide etc. is more concentrated in the lower portion of the atmosphere and decreases rapidly with increasing altitude. These weather elements absorb the outgoing long wave terrestrial radiation and thereby increasing the temperature of lower atmosphere than the upper layers. It means that the temperature decreases upwards because of decrease of absorption of terrestrial radiation with increasing height.

Thus, the heat is transferred to the atmosphere from the surface of the earth through the processes of convection, compression and radiation. As the altitude increases the amount of the heat transported upward decreases. Consequently every layer of air receives less heat than the layer lying below.

INVERSION OF TEMPERATURE

Inversion of temperature is an increase of temperature with height above the earth's surface. This is the reverse of the normal situation in which the temperature falls with height. It may occur from the surface of the earth or at altitudes far above the surface. A surface inversion is commonly experienced in hollows and valleys, especially in winter on calm and clear nights. The radiation causes considerable cooling and the cold air sunks down into these hollows and valleys, A pool of cold air thus lies there while on the mountain slopes above the air is warmer. The dense winter fogs of cities like London generally originate from this type of inversion.

DIURNAL RANGE OF TEMPERATURE

The variation of temperature during the day is known as the diurnal range of temperature or daily change of temperature. It is higher in inland areas where as low in coastal areas. During the night, due to the rotation, the earth is turned away from the sun, it is definitely cooler. Conversely, the temperature progressively rises when the sun is shining during the day time. Even during the day time there is variation of temperature. It is highest at noon, when the rays of the sun are vertically overhead while it is low in the morning and evening as the sun's rays are slanting at both the times. Though the insolation is most intense at noon, the highest temperature is recorded at 2 P.M. and the lowest temperature is recorded at 4 A.M. instead of mid-night. The clouds, winds precipitation and other factors like season nearness to sea and waters, latitudes, altitudes etc. have also significant influence and effects on the diurnal range of temperature. The difference between the daily maximum and minimum temperature gives the diurnal range of temperature.

SEASONAL RANGE OF TEMPERATURE

Besides its rotation around its own axis the earth also revolves round the sun once in a year. During its revolution the earth never changes its position in respect to its inclination. Thus, due to this inclination some portion of the earth comes nearer to the sun at one time while the other portion remains away from the sun. Hence, the temperature at different positions remains different. Every place has its hottest and coldest months. In the seasonal change of temperature the inclination of the earth determines the number of hours during which a particular place receives sunlight. This inclination is also responsible for the direct or slanting rays of the sun and insolation.

Besides this, many local factors like winds, currents, clouds precipitation, nearness to the sea, latitudes and altitudes also influence the seasonal variation of temperature. In the temperate and frigid zones the seasonal change of temperature is greatest. In the maritime place the hottest month is only warm while the coldest month is only cool as the second range of temperature is lowest in coastal areas.

ANNUAL RANGE OF TEMPERATURE

The earth occupies different positions in the course of its revolution round the sun. The earth remains inclined at an angle of 66.5° throughout its journey round the sun. The result is that the rays of the sun strike different areas of the earth vertically at different times of the year. The rays of the sun are overhead on the tropic of cancer 23.3 degree N. Latitude when the northern axis is most inclined towards the sun. In this position the Arctic circle receives sunlight for 24 hours and experiences six month's day while reverse is the position in the southern hemisphere. This happens on 21st of June when the Nr. Hemisphere enjoys the summer season and has the highest temperatures.

The conditions are exactly opposite when the southern end of the earth is most inclined towards the sun. The rays of the sun are overhead on the tropic of Capricorn 23.5 degree south latitude. In this position all the places within the Antarctic circle review sunlight for the full 24 hours. This position is reached on 22nd of December (winter season) when the earth has completed half of its journey round the sun. From March 21 to June 21 there is "spring" season in northern hemisphere. In the contrary from September 23 to December 21 there is "Autumn" in the northern hemisphere. In the northern hemisphere 21 June is the longest day while 22 December is the shortest day of the year. Thus, the annual range of temperature varies from month to month having the highest temperature in the month of June and lowest in the month of December every year. This all variation of temperature is due to the inclination of the axis of the earth.

ATMOSPHERIC PRESSURE

The pressure at a point due to the weight of the column of air above that point. The atmospheric pressure equals about 14.3 pounds per sq. inch at the sea level. With increasing height above sea level the overlying column of air being shortened the pressure decreases. The atmospheric pressure is registered in "millibars" normally measured by an instrument known as barometer.

Atmospheric pressure or air pressure is, thus, defined as total weight of a mass of column of air above per unit area. The unit area may be one sq. cm. or meter, one sq inch or foot. It is maximum near sea level i.e. 1034 grams per sq. cm. The weight of air present inside the human body exerts equal amount of outward pressure with which one does not feel such enormous weight of air on his head. Since the atmospheric pressure decreases with increasing altitude, therefore, there is an imbalance between outward air pressure of human body and inward atmospheric pressure. This imbalance result in nose and ear bleeding of a man at higher altitudes in the mountainous areas.

HORIZONTAL DISTRIBUTION OF AIR PRESSURE

The lines joining the places of equal atmospheric pressure are called isobars. The air pressure decreases with increasing height from sea level at the rate of 1 inch or 3.4 m per 600 ft. It varies from place to place and from time to time depending on temperature, altitude, water vaporous, winds and rotation of the earth etc.

(a) Temperature and Air Pressure

There is an inverse relationship between temperature and air pressure. It is very often said that if thermometer is high, barometer is low and vice versa. Hence, the general rule is that higher the temperature lower is the pressure and lower the temperature, higher is the pressure.

(b) Altitude and Air Pressure

The atmospheric pressure decreases with the altitude, because due to the compression the lower layers of air are thicker and heavier. On the other hand the air becomes thinner and lighter as we proceed higher up. It decreases with increasing altitude at the rate of one rich per 600 feet a.s.l.

(c) Water Vapours and Air Pressure

Water vapours are lighter in weight than the dry air. In other words the dry air is heavier than the air laden with water vapours.

VERTICAL DISTRIBUTION OF AIR PRESSURE

According to the general rule of high temperature belt with low pressure and low temperature belt with high pressure we can divide the earth into many belts.

(a) Equatorial Low Pressure Belt

The equatorial area is a low pressure belt as it is the hottest area. This belt extends upto 5 degree north and south latitude from the equator, which goes upto 20 degree north or south in summer. This belt is thermally induced as the ground is intensely heated by the direct and vertical sun rays which warms the lower layers of air by convention process. Warmed air, thus, expands, becomes light and rises upward results in low pressure.

(b) Sub Tropical High Pressure Belt

Another belt is the sub-tropical high pressure belt extending between 25 degree to 35 degree latitudes in both the hemispheres. The descent of winds results in the contraction of their volume and ultimately causes high pressure. That is why this zone is characterized by anticylonic conditions which cause atmospheric stability and aridity. The hot deserts of the world lie in this belt in the western parts of the continents.

(c) Sub-Polar Low Pressure Belt

The third belt is the sub-polar low pressure belt, which is located between 60 degree to 65 degree latitudes in both the hemispheres. There is low temperature throughout the year and as such there should have been high pressure. But in fact, the surface air spreads outward from this zone due to the rotation of the earth which results in low air pressure.

(d) Polar High Pressure Belt

The last belt is the polar high pressure belt, because of the prevalence of temperature below freezing point throughout the year.

The pressure belts seldom remain stationary in latitudinal zones as discussed above. Because there are daily, seasonal and annual changes in atmospheric pressure. These changes are due to northward and southward movement of the overhead sun and contrasting nature of heating and cooling of land and water. Thermal and dynamic are the two causes of formation of low and high pressure system. Thermal cause is that the hot air expands and becomes lighter, so it exerts low pressure which results in decreasing of pressure. The cold air on the other hand contracts and becomes heavier, so it exerts high pressure. Dynamic means physical power and forces producing motion. Hence, earth's rotation forces the equatorial and sub-polar air to move towards sub-tropical region and thuswise produce high pressure.[59]

WIND

Wind is a current of air, moving with any speed in any direction, but generally assumed to be parallel to the earth's surface. The direction of a wind is indicated by the point of the compass from which it blows e.g. a south wind is one which blows from the south. The speed of a wind is usually given in miles per hour on land and in knots at the sea surface. Winds may be classified according to their velocity and direction in which they blow. The velocity, speed and direction of the wind is measured by an instrument known as Anemometer.

Steep pressure gradient means a great difference between high and low pressure belts. Hence, the winds blow faster from high to low pressure areas. Conversely mild gradient means mild slope and winds are light. The movement of wind is directly controlled by the pressure of air. The direction in which the atmospheric pressure decreases is known as the 'barometric slope', and the rate of change in pressure is called "isobaric gradient". There are various factors governing the wind movement, some of which are as follows:

(a) Wind always blows from high pressure areas to low pressure areas. It means that it blows along the barometric slope.

(b) The earth rotates on its axis from west to east and due to this rotation the winds are deflected to opposite side i.e. east to west. In southern hemisphere reverse is the case.

(c) The isobaric gradient determines the speed and velocity of the wind. If these lines are close to each other the wind is stormy and if they are wide apart the wind is gentle or mild.

FERREL'S LAW

Prof. Ferrel, an American Scientist discovered a law about moving bodies such as winds, currents etc., which is known as Ferrel's law. According to this law the winds are deflected to their right in the northern hemisphere and left in southern hemisphere due to the daily rotation of the earth. Prof. Ferrel made experiments with spinning globe by allowing a hot tar to flow from the pole. He discovered that the deflection is the least at equator but deviation increases with increasing latitudes. The cyclones and anticyclones are the result of this deviation. The winds, are classified on the basis of their direction, speed and velocity. Some of the winds are as under which will be discussed under separate headings.

(1) Planetary Winds

These winds are constant and permanent winds. They blow throughout the year from one latitude to another. They always blow in the same direction. The three well known planetary winds are the trade winds, westerlies and Polar wind.

(a) Trade Winds

The trade winds blow from the sub-tropical belts of high pressure (30 degree north and south) and towards the equatorial region of low pressure, from the NE in the northern hemisphere and from the SE in the southern hemisphere. In many areas they blow with extreme regularity throughout the year, especially over the oceans. In continental interiors they blow much less steadily than over the oceans, but are fairly regular over the hot deserts. In the trade wind region the weather is normally fine and quite but tropical cyclones are often experience there.

(b) Westerlies

The westerly winds which blow with great frequency in regions lying on the poleward sides of the sub-tropical high pressure areas or "horse latitude". In winter they move southwards in the northern hemisphere affecting the Mediterranean regions bringing winter rains. They blow all the year round in temperate regions between 35 degree N to Arctice circle and 30 degree S to Antarctic circle. The winds derive their name from the prevailing direction. In the northern hemisphere this is mainly south-westerly and in the southern hemisphere north-westerly. The weather is marked by constant procession of depression and anticyclones moving eastward.

(c) Polar Winds

The winds blowing from polar high pressure belts towards the sub-polar pressure belts are known as polar winds.

(2) Periodical Winds

These winds are also known as seasonal winds. They change their directions according to the time or season. The well known periodical winds are monsoon winds and land breezes and sea breezes.

(a) Monsoon Winds

These are periodical or seasonal winds which blow from sea to land in summer and from land to sea in winter on a large scale. According to Prof. Flohn, monsoons are nothing but a modification of planetary wind system. He believes that in summer the equatorial belt of low pressure is shifted towards north and equatorial westerlies are embedded in tropical easterlies. Then these winds blow northward and are called "south east monsoons". As they move from the ocean to the land, they bring heavy rainfall. In winter, the equatorial belt of low pressure relates south-ward. The normal trade wind system is re-established which is termed as "winter monsoon".

(b) Land Breezes

The winds which blow from land to sea in the night are termed land breezes. As a matter of fact at night, the land gets cooled more quickly than the adjacent sea which remains relatively warmer. Hence, the air on the land is colder and heavier than the air on the adjacent sea. The pressure, therefore, is higher than that of sea. So the winds blow from land to sea during night, which are known as land breezes.

(c) Sea Breezes

The winds which blow from sea to land during the day time are termed as sea breezes. In the day, the land gets more heated than the adjacent sea and hence the low pressure develops there. On the other hand, the sea being cool develops higher pressure. Therefore, the winds blow from sea to the land, which are called as sea breezes. These land and sea breezes have significant effects on the weather and climate. Hence, the temperature remains uniform in the coastal regions and experience an equable type of climate.

(3) Local Winds

The winds blow locally are called as local winds. They develop due to the local difference in pressure and temperature. They blow in the lowest layers of troposophere. The local winds are of various types some of which are as under:

(a) Foehn or Chinook Winds

These are the hot and dry winds which occur in the middle latitudes. They descend on the leeward side of the mountains. These are called Foehn in Switzerland and Chinooks in USA and Canada. Since these winds are hot and dry they cause evaporation of soil moisture. Thus, the winds blown on the windward side of the mountain are moist and warm. They cause heavy rainfall on the leeward side of the mountain. Early ripening of grapes is due to chinook winds, besides removing of snow from the ground.

(b) Simoon Winds

These are also the hot and dry winds which blow during spring and summer in Arabia and Sahara deserts. These winds carry sand particles with them reducing the visibility. They erodes the rocks lying in its track. During the day time when they blow develop high temperature (more than 35 degree C) and very low pressure in a small areas.

(c) Karaburan

It is also a hot wind, which occurs in Tarim basin of Sinkiang. It blows during early spring when the interior of Asian land mass is intensely heated. It blows with high speed, so it sweeps up clouds of dust from the desert. It causes suffocation and discomfort to men and animals. It carries dust particles to long distances to deposit them as loess on the desert margins.

(d) Mistral

It is a cold and dry wind which blows from higher lands to the Mediterranean coast of France and moves through the Rhone valley. It blows with a high velocity of more than 60 km per hour. Hence the gardens and orchards are protected from it by thick hedges and cypress trees. It is as cold as below freezing point.

(e) Loo

Loo winds are hot and dry surface winds. They move in the northern plain of India during summer season especially in the months of May and June. They blow from west to east and eause loss of lives in the plains of Bihar and U.P.

(f) Katabatic Winds

They are also known as mountain breezes or gravity winds. At night the hill tops and plateaus lose heat rapidly by radiation and become cold. The air in contact with them also becomes cold and heavy and blows down the valley.

(g) Anabatic Winds

In mountainous areas the hill tops and plateaus are heated during the day time, which warms the air above them. The warm air expands, becomes light and rises up. The winds from the low land blows up the valley sides to take its place which is called as valley breezes or anabatic winds.

13
Climatic Classification

Climate is the average weather conditions of a place or a region throughout the seasons. It is made up of so many elements such as wind, rainfall, humidity and temperature. Some of these elements are highly variable from day to day and season to season. The climate is also governed by different factors such as latitude, position relative to continents and oceans, prevailing winds, ocean currents, vegetation cover and distance from the sun. The locial climates are still further modified by altitude, proximity of mountains etc. Thus it is difficult rather impossible to determine upon any system of classification of climates that will be both simple and accurate.

However, there are numerous accepted classification of climates having qualitative or quantitative criteria, or both, for the delimitation of its climates. Because the criteria of delimitations are different in the various classifications, some of them accomplish one thing better than others do. The Koeppen Geiger's classification is the one which seems to be most universally accepted. The Koeppen system of classification is largely quantitative. It is principally based upon the amount of precipitation and temperature. These elements, considered separately and together, constitute the framework for the divisions of climate.[64]

KOEPPEN'S CLASSIFICATION

Wladimir Koeppen (1846 - 1940) a German botanist and climatologist presented his classification first in 1900 on the basis of vegetation zones. He revised it in 1918 on the basis of temperature and precipitation. He again modified his scheme in

1931 and 1936. It was again modified in 1953 by another scientist Geigger-Phil, thus published as Koeppen-Geigger's classification of climate. As a matter of fact Koeppen divided world climates into five primary or major categories and designated them by capital letters of A,B,C,D, and E. The temperatures characteristics of each of the primary categories are as follows:

A: Humid tropical climates with winter less seasons, warm and worst conditions throughout the year with mean temperature above 18 degree C.

B: Dry climates, where evaporation exceeds precipitation and constant water deficit throughout the year.

C: Humid mesothermal or middle latitudes warm, temperate climates having mild winters. The average temperature of the coldest and warmest months being 18 degree C and below 18 degree C respectively.

D: Humid microthermal or cold forest climates with severe winters. The average temperatures of coldest and warmest months being 3 degree C and above 10 degree C respectively.

E: Arctic or polar climates with summer less season having an average temperature of the warmest month below 10 degree C.

Each of the above primary cartegories is sub-divided into two or more secondary categories represented by small letters as follows:

f: Precipitation throughout the year with a minimum of 6 cm in each month. The average temperature of the coldest month being more than 18 degree C.

m: Monsoon climates has excessively heavy precipitation, short dry season with average precipitation in driest month less than 6 cm.

w: Winter dry season, driest month has less than 2.4 inches mean precipitation.

h: From German word "heiss" meaning hot with mean annual temperature above 20 degree C.

k: From German word 'kalt' meaning cold with temperature below 18 degree C.

a: Hot summer with temperature above 20 degree C.

b: Warm summer (15 degree C).

c: Cool summer.

d: Severe Winter (-38 degree C).

s: Well defined summer dry season, with mean precipitation for the driest month of the summer half year less than 1.6 inches and less than one third that of the wettest month of the winter half year.

According to Koeppen's classification the major 5 primary divisions already discussed are further sub-divided into 11 sub-types on the basis of seasonal regimes.

(1) Tropical Rainy Climates or 'A' Climate

On the basis of periodicity and regime of precipitation this type has been further divided into 3 types as under:

(i) Af: Climate

All month hot and moist, humid tropical climate, precipitation in driest month more than 6 cm. The seasonal distribution of precipitation is more or less uniform with very low daily range of temperature.

(ii) Aw: Climate

All months hot with summer rains and winter droughts. Precipitation of at least one month less than 6 cm with high temperature through out the year.

(iii) Am: Climate

All months hot, some months excessively moist. Monsoon climate with short dry season but sufficient annual precipitation with dense forests.

(2) Dry Climate or 'B' Climate

On the basis of annual temperature and precipitation this type of climate is sub-divided into a four subtypes as under:

(i) Bsh: Climate

Hot and semi-arid. Tropical steppe climate with mean annual temperature of above 18 degree C.

(ii) Bsk: Climate

Semi-arid, cold winters. Middle latitude cold steppe with annual temperature of below 18 degree C.

(iii) Bwh: Climate

Hot and arid (Hot desert) Tropical desert climate with average annual temperature of more than 18 degree C.

(iv) Bwk: Climate

Cold deserts with severe winters. Middle latitude cold desert climate with mean annual temperature of below 18 degree C.

(3) Humid Mesothermal or Warm Temperate Rainy Climate or 'C' Climate

This 'C' type of climate has been sub-divided into following types:

(i) Cfa: Climate

Hot summers, mild winters; all months moist.

(ii) Cfb: Climate

Warm summers, mild winters, all months mild.

(iii) Cfc: Climate

Cool summers, mild winters; all months mild. Average temperature below 22 degree C.

(iv) Cwa: Climate

Hot, moist summers, mild, dry winter. Warmest months above 22 degree C.

(v) Cwb: Climate

Short warm, mild summers, mild, dry winters.

(vi) Csa: Climate

Mild, moist winters, hot, dry summers.

(vii) Csb: Climate

Mild, moist winters; warm dry summers.

(4) Humid Microthermal or Cold Forest or 'D' Climate.

The 'D' type climate has been further divided into 8 sub-divisions as under:

(i) Dfa: Climate

Severs cold winters, hot long summers; all months moist.

(ii) Dfb: Climate

Cold winters, warm short summers; all months moist.

(iii) Dfc: Climate

Cold winters, short, cool summers; all months moist.

(iv) Dfd: Climate

Cold winters, short, cool summers; all months moist.

(v) Dwa: Climate

Cold, dry winters; hot moist summers.

(vi) Dwb: Climate

Cold, dry winters; warm moist summers.

(vii) Dwc: Climate

Cold, dry winters; short cool and moist summers.

(5) **Polar Climate or 'E' Climate**

Polar or arctic climate or E Type climate is further sub-divided into 2 sub-types as under:

(i) Et: Climate

Tundra climate; very short and cool growing season. The temperature of the warmest month being below 10 degree C but above 0 degree C.

(ii) Ef: Climate

Permanent snowfields, temperature of all months below freezing point (-0 degree C), icecap or perpetual frost climate.

MERITS

W.Koeppen, being basically a botanist, has propounded his classification on the basis of two easily measurable weather elements e.g. temperature and precipitation. As a matter of fact both these elements are most widely and frequently used effective weather elements which control the climate. His scheme is primarily based on the relationship between floral types and their characteristics and climatic characteristics of a given place. He paid due attention to the loss of moisture through evaporation, as he included precipitation. He says that in fact it is not the total annual precipitation which matters, rather it is the effective amount of water which is available to plants or flora. This scheme is more descriptive, generalized and simple, hence it is universally accepted.

DEMERITS

Inspite of several merits discussed above, there are many drawbacks and objections to the Koeppen system. One criticism is that the various numerical criteria are arbitrary and that there is apparently no justification for the selection of only some of them. He gave undue importance to temperature and precipitation in his scheme neglecting all other weather elements. He also ignored the consideration of causative factors of climate, such as airmasses, winds, daily range of temperature etc. With his division, sub-division and subtypes by letter symbols to indicate climate types, he made his scheme very difficult to memories for the students.

CLASSIFICATION BY THORNTHWAITE

C. Warren Thornthwaite, an American climatologist presented the classification of climates of N. America in 1931. Latter he extended it to the classification of world climates in 1933. He further modified his scheme and published the revised and detailed classification of world climates in 1948.

(1) 1931 Classification

Thornthwaite also gave more consideration to natural vegetation as the major indicator of climate of a place. He took the amount of precipitation and temperature as having the paramount control on vegetation. He used two factors for the delimitation of boundaries of different climatic regions as follows:

(a) Precipitation Effectiveness

It is only that amount of total annual precipitation which is available for the growth of flora (vegetation). He identified five humidity zones of vegetation on the basis of precipitation efficiency index (P/E Index).

P/E Index Rainfall in Inches	*Humidity Zones*	*Vegetation*
127	A (Wet)	Rainforest
64-127	B (Humid)	Forest
32-127	C (Sub-Humid)	Greenland
16-31	D (Semi-Arid)	Steppe
16	E (Arid)	Desert

Thronthwaite give four sub-divisions of each of the above humidity zones on the basis of seasonal distribution as follows:

Ar;	As;	Aw;	and	Ad:
Br;	Bs;	Bw;	and	Bd;
Cr;	Cs;	Cw;	and	Cd;
Dr;	Ds;	Dw;	and	Dd;
Er;	Es;	Ew;	and	Ed;

where in

r - adequate rainfall in all seasons.

s - rainfall deficient in summer.

w - rainfall deficient in winter.

d - rainfall deficient in all seasons.

(b) Thermal or Temperature Effectiveness

He believed that temperature had important contribution in the growth of vegetation. Thus, on the basis of temperature effectiveness index (T/E Index) Thronthwaite divided the whole world into 6 temperate zone as follows:

Temperate zones	*T/E Index Rainfall in Inches*
A. Tropical	127
B. Mesothermal	64-127
C. Microthermal	32-63
D. Taiga	16-31
E. Tundra	1-15
F. Frost	0

2. 1948 Classification

Thornthwaite modified his old scheme and presented a more detailed and elaborated classification of climate in 1948. He based his new scheme on the concept of potential evapotranspiration (PE). It may be pointed out that PE is calculated from the mean monthly temperature with day length.

Thornthwaite developed four indices to determine the climate of a place e.g. (a) moisture index, (b) potential evapotraspiration of thermal efficiency index, (c) aridity and humidity indices, and (d) concentration of thermal efficiency.

MERITS

(1) Like Koeppen's scheme his scheme is also empirical as well as quantitative as the boundaries of different climates are determined on the basis of quantitative parameters derived from precipitation and temperature.

(2) Vegetation is made as the basis for the identification of climatic zones.

(3) Various letter combinations are used to designate different climatic types etc.

(4) This scheme yield the number of climatic types (32) than Koeppen's climatic types.

DEMERITS

(1) On the basis of above large number of types the Thornthwaite's classification system becomes so complex that it is difficult rather impossible to memorise it.

(2) This scheme also suffers from a serious problem of non-availability of the data of evaporation for all the places. Thus, the lack of adequate climatic, data makes it difficult for the precise demarcation of climatic boundaries of the world. For these reasons the Thornthwaite's classification of climatic scheme could not get more popularity and worldwide recognition.

TREWARTHA'S CLASSIFICATION

G.T. Trewartha, an American climatologist after making several revisions and modifications in the Koeppen's classification presented his simple scheme of climatic classification of the world. Thus, his scheme is a compromise between purely empirical and genetic methods of climatic classification. He also opposed to produce large number of climatic types and complex schemes such as presented by Koeppen and Thornthwaite. He identified only six major climatic types as A,B,C,D,E, and F. Out of these B climates were determined on precipitation criteria while others were determined on the basis of temperature criteria.

Trewartha's classification is very simple, unambiguous and a mixture of both empirical and genetic method of climatic classification. It uses only precipitation and temperature and avoids the vigorous statistical calculations. He also considered the effects of land and water surface on the climate of a place. Hence, due to its simplicity the classification became more popular among the readers.

WORLD'S MAJOR CLIMATIC TYPES

(1) Tropical Rain Forest

It is popularly known as equatorial type which occurs in a belt between 0 degree to 10 degree north and south but at some places it extends upto 25 degree north and south latitudes. The sun's rays fall perpendicularly, so the temperature is high (about 27 degree C) all the year round. Heavy convectional precipitation results, with trade winds converging and rising in this belt, called doldrums. The rain falls almost every day after about 3 PM with an annual rainfall of 250 cm. The nights are called the winters of equatorial region, as there is no winter season. As the season is always wet and hot, the region is characterised by very luxuriant growth of vegetation. The forests consist of tall hard-wood trees which remain evergreen.

It includes the Amazon basin, Congo basin, Zoire basin, Indonesia and south east Asia. The people are generally backward with main occupation as hunting in these basins. Singapore, Colombia and Para are important port towns of this region. The chief products are cacao, coconut, palm, bananas, rice, millets, petroleum, tin, copper etc.

(2) Monsoon Climate

This climate is dominated by monsoon winds. It rains mostly in summer for about four months brought by south west monsoon winds blowing from sea to land with an annual rainfall of about 300 cm. In winter the winds blow from land to sea so they are dry. The annual average temperature is about 26 degree C. The rainfall determines the vegetation so the luxuriant evergreen forests are generally found in the belt. Sal, Sesam and Teak being the most important trees besides other deciduous trees. The secondary most important products are oils, gums, resins and lac etc. The elephant, buffalo and cattles are the common animals of this region, besides some flesh eating wild animals. The main occupation of the people is agriculture, which supports one-fourth of the population of the world in this region. The region is overlaid with rich alluvial flood soil renewed every year. The main crops being rice, wheat, cotton, sugarcane, oilseeds, jute and tea. The region includes western Guiana (Africa), eastern Brazil, West Indies, south east Asia, Coast of Columbia, North Australia and eastern Amazon valley.

(3) Tropical Grasslands

The climate is typically found in tropical grasslands between 5 degree and 15 degree north and south. It is a transitional zone between tropical rain forests and the hot deserts. The annual rainfall is 160 cm, mostly in summer, which is uncertain and variable. The annual average temperature remains 23 degree C. It includes the grasslands of S. America, Savannas of Africa. Parkland of Australia and Campas of Brazil. The natural vegetation of Savanna is tall grass (10′) with scattered trees. On the whole these regions are still undeveloped and backward.

(4) Tropical and Subtropical Desert

The climate of this region is hot and dry dominated by subtropical anticyclones. The average annual rainfall is about 25 to 40 cm. Under clear skies the evaporation is 20 times more than the precipitation. The average annual temperature is 38 degree C. The belt is situated near tropics between 20 degree and 30 degree north and south latitudes in the west of the continents. It includes the deserts of Sahra (N. Africa), Thar (India), Arabian desert, south west USA, south western Africa and Australian (Great Sandy and Victoria) deserts. The climate is not conducive for vegetation growth, because of extreme aridity, high temperature and high rate of evaporation.

(5) Tropical and Subtropical Steppe Type

It is also called as temperate grassland type of climate. It is found in continental areas between 30 degree to 50 degree latitudes. It includes steppe areas of USSR, Priaries of N. America, Pampas in S. America, Veld in S.Africa and Downs in Australia. The annual rainfall is 30 cm while the annual mean temperature is 21 degree C. The mountain barriers restrict maritime influence.

(6) Mediterranean Type

This type of climate experiences hot summer and wet, mild winters. The average annual rainfall varies from 40 to 60 cms, most of which is received during winters, mostly related to cyclones The annual average temperature is 16 degree C. This climate occurs in western parts between 30 degree to 45 degree latitudes. The principal countries of this region are Portugal, Spain, S. France Italy, Greece, Turkey, Syria, Israel, Algeria, Morocco, California, Chile and south eastern and western Australia.

As its name indicates, it represents the climate that prevails around the Mediterranean sea. It is a transition zone between the southern hot deserts and northern cool temperate regions. On the wetter parts of higher altitudes the forests are found with pine, chestnut, mulberry, fruit and evergreen trees. The main occupation of the people is agriculture besides dairy farming, cattle rearing etc.

(7) China Type

This type of climatic region lies in the east of warm temperature belt lying between 30 degree and 45 degree north and south latitudes. The annual average temperature is 19 degree C and annual rainfall of 120 cm with a summer maximum. The weather is variable due to Hurricanes and Typhoons. This type of climate is peculiar to China, hence known as China type. It occurs is China, south eastern USA, Argentina, S. Brazil, Japan and eastern Australia

(8) N.W. European Type

This region has warm summers and cool winters with well distributed rainfall throughout the year. The annual average rainfall is about 140 cm. The westerlies are onshore here so they bring heavy rains. The cyclones are common, hence, the weather is variable and uncertain. The annual average temperature is 10 degree C. This type of climatic region lies between 25 degree to 60 degree north and south latitudes in the western margins. It occurs in north west Europe, British Isles, Coastal N. America, S. America and Australia.

(9) St. Lawrence Type

This region is the eastern margins of the cool temperate belt which lies between 45 degree and 60 degrees north latitudes. It concurs in eastern. Canada (St. Lawrence basin), north east USA, south east. Siberia, Manchuria and northern Japan. It has cold winters and warm, wet summers. The rainfall is well distributed but more in summers. The summer temperature is 20 degree C while in winters it remains below zero degree.

(10) Taiga Type

In this belt the winters are extremely cold and the ground remains covered with snow and ice for about 6 months. The average temperature is 5 degree C. The climate is dominated by polar and arctic air masses. The annual rainfall is 50 cm. This region lies between 50 degree and 661 degree north latitudes. This Siberian type of climatic region includes Siberia, northern Europe, Finland, Scandinavia and Canada.

(11) Tundra Type

This region is bordering the Arctic Ocean, including the far northern lowlands of Eurasia, Canada, Alaska and Green Land. In southern hemisphere it is found in Antarctica. This region remains under snow and ice for about 9 months. The region is often called as cold desert. The Summer temperature is 10 degree C while in winters it is below Zero degree with annual rainfall of less than 30 cm.

(12) High—Land Climate

This type includes the Tibetan plateau, the Himalayas, the

Rockies and the Andes mountains. This region is characterised by the vertical gradient in all the principal climatic elements. It experiences high insolation, low temperature, low air pressure, large diurnal range of temperature and large rainfall with snowfall at higher altitudes.

ROLE OF CLIMATOLOGY IN DAY-TO-DAY LIFE

The basic needs of man and the chief factors in man's life are his food, shelter, clothing, habits, economic activities and his movement from one place to another. The role of climatology in day to day life of a man will be more clear from the following factors:

(a) Food

In several parts of the world people are vegetraian, while in other parts they depend on meat diet. They eat flesh of cattle, sheep, goat and fish. Because they need more heat and energy for their body to fight against the cold, which is directly one of the climatic factor. The animals are dependent upon vegetable Kingdom, which is again directly controlled by the factors of climate i.e. rainfall, temperature, humidity and wind etc. It has been observed that people depend on animal food in regions of either heavy rainfall or totally scanty rainfall (desert with no vegetation), where temperature is too low to grow any kind of vegetation.

(b) Clothing and Shelter

Clothing and shelter of a man is also directly determined by the climatic factors. In high temperature and heavy rainfall regions there are dense forests (Equatorial regions). The people remain half naked there round the year. They built their houses on the branches of trees as in Amazon (South America) and Congo basin (Africa). But in very cold regions like Tundras (snow bounded N. America and N. European region), man wears tight clothes of animal skin and built their houses with narrow passages, so that no cold air enters the house. The light dress of Indian, and the tight dresses (close fitting) of Europeans is directly controlled by the climatic factors.

(c) Habits

Even the habits of man like talking, walking, the way of taking their food, living outdoor or liking indoor life all depend upon the climatic factors of a place, known as the role of climatology in day to day life. In cooler countries or colder regions, people take their food after getting full dressed while in warmer regions with least dress on.

The people generally like outdoor activities in warmer areas but on the other hand they remain indoor and love club life in colder regions. Again the people of colder areas are generally advanced and hard working because they struggle hard for their existence, than the people of hot and moist areas, who become ease loving, slothful and lazy. The people of warmer areas have fluency of language and higher tone as compared to the people of colder regions who talk in slow motion and low tone.

Availability of even raw-material for industrialisation depends upon climatology. Even the quality of wool (Sheep) depends on the climaic factors like that of temperature and rainfall. Sheep of cool and dry climate yield better fleece (wool) than the sheep of warm and moist areas. The minerals and metals in present use are also the results of past climates.

(d) Population and Migration

Population and movement of people also depends on climate of a place. The plain temperate regions are thickly populated than equatorial regions and tropical grasslands. The monsoon lands and their river valleys support a large population than Amazon and Congo basin and desert areas of the world. The migration of people from one place to another is also the result of climate to certain extent. The last ice age was responsible for the migration of Aryans from Central Asia to another parts of the world, particularly to (India and Pakistan) southern Asia.

(e) Physical Features of Man

The colour of skin, features of the face and even he height and race of a man depends on climatology. In hot and moist region and colour of the people is generally red-brown. In warm and dry regions yellow, in colour moist regions white and olive and in hot regions (Africa) black tanned people have been observed. The people of colder places usually have longer nose, so that the inhaled air gets time to warm, while the people of comparatively warmer places have very short noses. In short, as a matter of fact, the climate has a great influence on the human activities. The use of different types of clothes in different regions, variety in house building and food stuffs are all related to climatic conditions. Natural vegetation and cultivated crops are also related to climate. Climate sets certain limits to the range of crops that can be cultivated in a particular area. Each climatic region is best suited for a particular sets of crops. The different climates necessitate the exchange of products among the different countries and thus provide a basis for international trade. The mode of life of the people is also related to the climatic conditions[65]. Hence the role of climatology in day-to-day life of man is very important in every respect.

14
Air Masses and Fronts

INTRODUCTION

A homogeneous mass of air covering a considerable area of the earth's surface and bounded by frontal surfaces is known as an air mass. Air masses are usually classified according to the regions from where they originate, e.g. tropical or polar. According to the weather they are of maritime or continental type. Thus, it may be classified, for example, as tropical maritime or polar continental. According to Barry and Chorley (1968), "an air mass is a large body of air whose physical properties, especially temperature, moisture contents and lapse rate are horizontally uniform for hundreds of kms." While A.N. Strahler defines an air mass as "a body of air in which the upward gradients of temperature and moisture are fairly uniform over a large area".

The air mass is different from that of wind, which blows horizontally along the earth's surface, whereas an air mass always moves vertically when it is unstable. An air mass has two basic properties (1) has vertical distribution of temperature and (2) has moisture or water vapour contents. Both these properties of air mass affect the weather of the region on which it tranvels. The place from which an air mass is originated is termed as source region. They are the extensive areas on the surface of the earth. The air masses affect the weather conditions and are related to atmospheric disturbances, formation of fronts and transference of heat from lower to higher latitudes. The air masses can be classified on two main basis i.e. source region of the air masses and modification. On the basis of source region, air masses are classified into (1) Maritme air mass which is further sub divided into (a) polar maritime and (b) maritime tropical air Mass. (2) Continental air mass, sub-divided into (a) continental tropical and (b) polar continental air mass.

The air masses are modified as they move away from their source regions in two ways e.g. by thermodynamic and dynamic origin which develops between warm moist tropical air mass and cold dry polar air mass. Thuswise, its formation is closely related with the air masses. The plane which separates the two air masses is termed as front. Two types of fronts are identified in the temperate cyclone: (1) The warm front and (2) cold front. The warm front has a gentle slope and warm air ascends gradually along the front over cold air below. Cold front has steeper slope and warm air is pushed rapidly by the advancing cold air. Ascent of warm air along the fronts leads to rainfall.

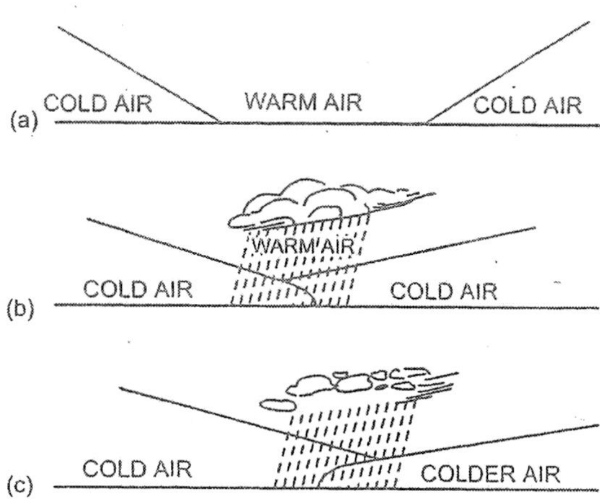

Fig. 1. (a) Cold and warm fronts; (b) cold type occlusion; (c) warm type occlusion.

PROPERTIES OF WARM FRONT

(a) It is pushed by warm air mass behind it. It is also associated with heavy rainfall.

(b) When the warmer air mass is more active it moves towards the cold air mass pushing its front over the wedge of the cold air mass.

(c) It is relatively unstable than cold air front.

(d) Warm front always overrides the cold front whenever they conflict with each other.

(e) It is always found in a straight line. Its vertical movement is more than cold front.

(f) It horizontal movement is slower than that of cold front.

PROPERTIES OF COLD FRONT

(a) Cold front is more stable than warm air front. It is pushed by the cold air mass behind.

(b) It always pushes the warm air front upwards leading to condensation.

(c) Its horizontal movement is more than that of the warm front.

(d) Vertical movement of cold front is very less as compared to warm front.

(e) Sometimes cold air override the warm air leading to violent convectional overturning.

When the two fronts (cold and warm) meet each other, either the warm air front overrides the cold front or cold front under runs the warm front. The line along which they mix experiences a change in temperature. This part of the front is known as occlusion front. It is highly unstable, associated with heavy rainfall, thunder storms, thunder showers and hail storms. If the overtaking cold air is colder than in the front, the occlusion is known as cold occlusion otherwise it is warm occlusion.[60]

ATMOSPHERIC DISTURBANCES

Atmospheric disturbances occur due to certain local interferences as local conditions produce irregularities in the atmosphere which cause variation in the air pressure. Due to this variation of pressure, the air begins to move and is known as variable winds. These are generally irregular whirlwinds or terrific storm. Their appearance is usually quite sudden due to extraordinary causes. These horizontally and vertically moving air or variable winds are known as cyclones and anticyclones. They

Table 1.

Sources and Characteristics of the Principal Air Masses

S.No.	Type	Symbol	Origin	Temperature and Moisture Characteristic of Sources	
				Winter	*Summer*
1.	Polar continental	cP	Alaska, Canada	Very Cold, Dry	Cool Dry
2.	Polar Maritime	mP	Nr. Atlantic Pacific	Moderately Cold, Humid	Cool, Humid
3.	Polar Pacific	pP	Nr. Pacific	—	—
4.	Polar Atlantic	aP	Nr. Atlantic	—	—
5.	Tropical Continental	cT	SWn. USA Nr. Mexico	—	Hot, Dry
6.	Tropical Maritime	mT	Subtropical Waters	Warm Humid	Hot, humid
7.	Tropical Pacific	pT	Subtropical North Pacific	—	—
8.	Tropical Atlantic	aT	Subtropical N. Atlantic	—	—
9.	Tropical Gulf	gT	Gulf of Mexico and Caribbean	—	—

blow spirally either towards a low pressure centre known as cyclone or towards high pressure centre called an anticylone.

CYCLONE

A region in which the atmospheric pressure is low as compared with that of adjacent areas is known as cyclone. In a cyclone the wind is spirally moving in towards a low pressure centre with a high pressure on the surrounding sides. The wind movement is anti-clockwise in the northern hemisphere and clockwise in the southern hemisphere. Cyclones are also termed as atmospheric disturbances. They range in shape from circular, elliptical to vee-shape. When the winds attain gale force by high velocity, they are known as cyclonic storm. From the locational point of view cyclones are classified into two main types e.g. (1) temperate cyclones or extratropical cyclones and (2) tropical cyclones.

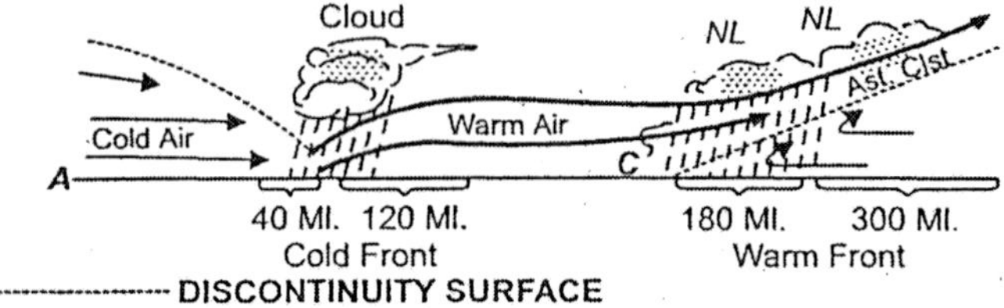

Fig. 2. Vertical section of a cyclone.

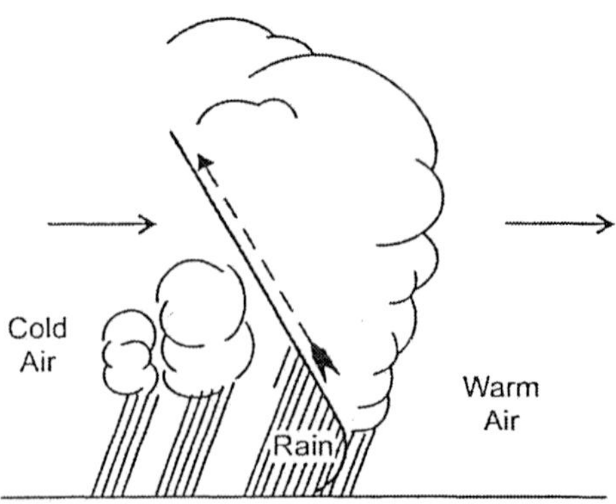

Fig. 3. Upward movement of warm air at a Cold From (broken arrow), and direction of movement of the front.

Anti-Cyclone

A region in which the atmospheric pressure is high as compared with that of adjacent areas is known as anticyclone. It has a high pressure centre hence the wind movement is clockwise in the northern hemisphere while anticlockwise in the southern hemisphere. The wind is spirally moving outward. In the temporary anticylones of temperate latitudes, quite and settled weather conditions are the characteristics in contrast to the weather of cyclones (depressions). In summer, skies are often cloudless and temperatures are relatively high, but in winter there is so much radiation that the lower layers of air become excessively cooled which results in general fog. Such conditions may persist for a considerable period as they often remain almost stationary for several days and move very slowly.

It has been often noted that in between two cyclones exists an anticyclone. As they move with the prevailing winds in the same direction, they set up a sort of procession of highs and lows. They are common in temperate and torried zones but dominate in tropics during summer and in temperate zone during winter. Anticyclones move in accordance with the direction of winds. They move with trade winds from east to west in the tropics and with westerlies from west to east in the temperate zone.

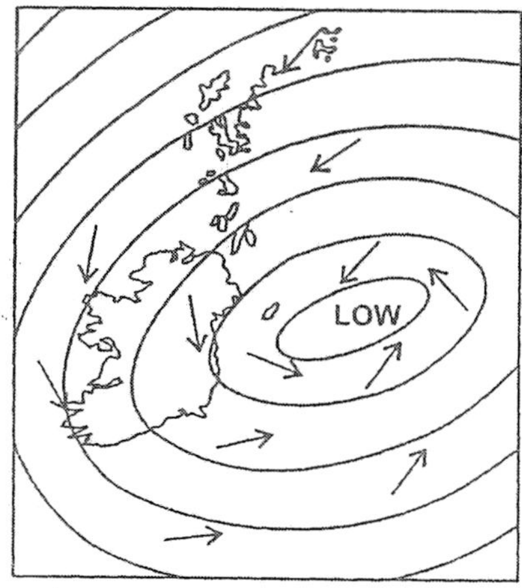

Fig. 4. A Cyclone.

TYPES OF ANTI-CYCLONES

There are two separate classifications according to Hanzlik and Humphreys. Hanzilk put forth two distinct types as follows:

(1) Cold Anti-cyclone originates in very cold region and are the counterparts of cyclones. Due to the intense cold a high pressure area develops. Siberia and Central Asia are classic examples, where such anti-cyclones often develop.

(2) Warm Anti-cyclone develops near sub-tropical highs as the descending air gets heated and dry. It gets compressed and pressure increases due to compression. These are usually of transitory character.

Fig. 5. Section through a Cold Front Occlusion. Arrows show direction of movement.

Humpreys divided anti-cyclones according to their origin into three types:

(1) The Westerlies try to rise up the bulge over to the equator by centrifugal force and the trade winds try to get over to the poles because of the less gravitational pull. It results into a mechanical squeeze in the region between 30 degree to 35 degree latitudes, making it a high pressure centre especially over the oceans. In this way the mechanical anti- cyclones are originated.

(2) The high pressure is the common feature of Greenland and Antarctica because of the higher latitudes and free radiation. Hence these regions are the permanent centers of anti-cyclones known as radiational anti-cyclones.

(3) The anti-cyclones resulting from low temperature and high pressure in oceanic areas having semi-permanent

nature are called as thermal anti-cyclones. The temperature of the water is reduced by cold water currents of the oceans which promotes high pressure centers.

WEATHER CONDITIONS

An anti-cyclone is associated with different weather conditions. The general weather associated with it is fine and dry. In its middle the weather largely depends upon the season and on the margins the weather conditions depend upon the source of the wind. The sky remains generally clear except for some hazy appearance over the horizon. The days are hot in summers (calm and fine weather) but in winters foggy conditions are common particularly at nights (fogs and frosts).

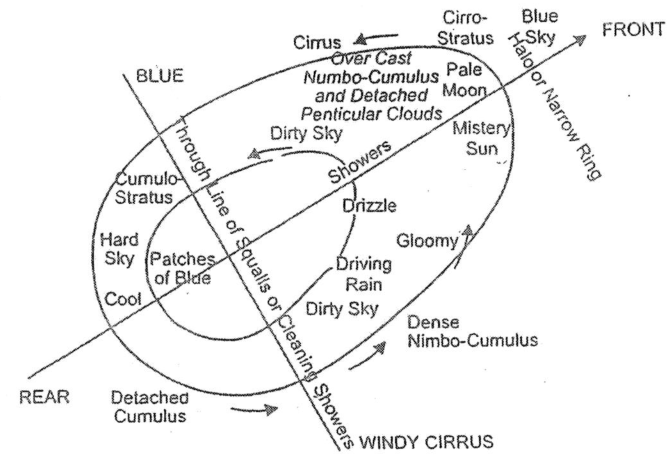

Fig. 6. Weather within a cyclone.

TROPICAL CYCLONES

A region of low atmospheric pressure, relatively small in area but accompanied by violent storm conditions, which originate in the tropics, are called tropical cyclones. They originate in the area of the 'Doldrums' where they are well beyond the equator and which generally travel westward with the prevailing winds of the area. They are known to occur in all the tropical seas near the equator, except in the Atlantic, south of the equator. In the region of the West Indies, Mexico, Central America and Australia they are reffered to as 'Hurricanes'. In the regions of western Pacific from Indo-China to Japan they are called as 'Typhoons' while in the Indian Ocean, they are reffered to simply as Cyclones. On a weather map, the isobars are more cicurlar and closer together, which means the pressure gradient is much steeper than in cyclones of middle latitudes. This pressure gradient accounts for the high wind velocities and consequent destructiveness of tropical cyclones.

Fig. 7. Upward movement of warm air at a Warm Front (broken arrow) and direction of movement of the front.

ORIGIN

The tropical cyclones develop due to the local convection currents acquiring a whirling motion by the rotation of the earth. The conditions most suited for their origin and development are:

(1) Continuous supply of abundant warm and moist air. Tropical cyclones originate over warm oceans (27 degree C) during summers.

(2) It needs higher value of coriolis force. It is caused due to the deflection in wind direction which is the result of coriolis force.

(3) Majority of tropical Cyclone originate within a belt of 5 degree to 20 degree latitudes in the western parts of oceans. They extend to 30 degree latitudes in summer season.

(4) Highly saturated atmosphere and great heat is favourable for its origin.

These conditions are usually found in doldrums or equatorial calms especially over the ocean surface between 5 to 15 degree latitudes in both the hemispheres. They influence the weather conditions of coastal areas of the continents.

Shape and Size

The shape and size of tropical cyclones varies considerably. On an average their diameters range between 80 to 300 km but sometimes they are as small as 50 km or even less. They are generally cicrular or elliptical in shape having a length ratio of 2:3.

Velocity and Speed

The tropical cyclones advance with varying velocitities. Some of them move at a speed of 32 km per hour while hurricanes attain a velocity of more than 180 km per hour. Tropical cyclones become more vigorous and move with a terrific speed over the oceans but on land areas they become feeble and their speed is retarted to zero in the interior of the continents. This is the main reason, why these cyclones affect only coastal areas like south and south east coasts of USA, Tamil Nadu, Orissa, West Bengal and Bangladesh.

Weather Associated

The weather associated with tropical cyclone is the increase in air temperature, wind velocity, decrease in air pressure, appearance of cirrus and stratus clouds and emergence of high waves in the oceans. The clouds are thickened and become cumulonimbus which yield heavy rains, with thunder and lighting. A single storm yields about 250 mm of rainfall and sometimes by relief barrier about 800 mm. The sky is overcast with thick and dark clouds making very poor visibility for a few hours. The centre is known as the eye of the tropical cyclone. This eye is characterised by a small patch of blue or clear sky and there is an extent of calm and dry weather. The weather suddenly changes with the arrival of the rear portion of the cyclone. There is heavy down pour with violent thunderstorm and gusts of winds. The weather become again dull and muggy which passes away as the tail of the cyclone is reached.

Types

Generally the tropical cyclones are divided into two principal types and four sub-types on the basis of intensity e.g. (1) Weak cyclones, sub-divides into (a) tropical disturbances (b) tropical depressions and (2) strong and furious cyclones, sub-divided into (a) hurricanes or typhoons (b) tornadoes or troprical storm.

GENERAL CHARACTERISTICS

Cyclone developed in the tropical regions of Cancer and Capricorn are called tropical cyclones. These are not regular and uniform like temperate cyclones. There are numerous forms of these cyclones which very considerably in shape, size, velocity and weather conditions. They are characterized by the following salient features:

(a) The centre of the cyclone is characterized by extremely low pressure. Isobars are more or less circular but are fewer in number. That is why winds hurriedly rush up towards the centre and attain gale velocity.

(b) The tropical cyclones are not characterized by temperature variation in their different parts as they do not have different fonts (warm and cold) as compared to temperature cyclones.

(c) There are no different rainfall cells in it as in the case of temperate cyclones, hence each part yield rainfall.

(d) They are not always in motion. They become stationary over some places for several days and give heavy rainfall causing floods and become disastrous natural hazards, because of their high wind speed.

(e) Its tracks varies in different parts. Due to trade winds they move from east to west upto 15 degree latitude and poleward between 15 degree to 30 degree and then again move westerly. In subtropical regions they become very weak.

(f) They are confined to a particular period of year, mainly during summer season.

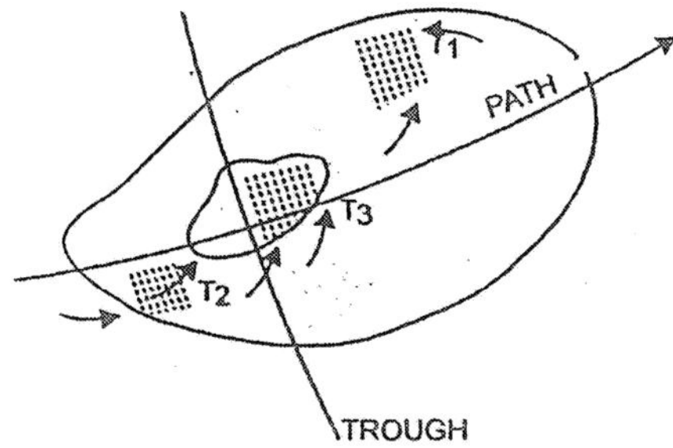

Fig. 8. Weather in a strongly developed Temperate Cyclone.

Temperate Cyclones

Temperate cyclone or wave-cyclones are the depressions having low pressure in the centre and high pressure outward. They originate in the middle latitudes by covering and rising air, cloudiness and precipitation. They are also known as lows, troughs and depressions owing to their isobar as V-shaped. Their dominance is between 35 degree to 65 degree latitudes in both the hemisphere. The polar fronts created due to opposing air masses i.e. tropical (warm) and polar (cold) are responsible for the origin and development of temperate cyclones. These cyclones are most dominant over the North Atlantic Ocean especially in winter season. There is a marked difference of temperature and rainfall between their front and back. After their formation, they move with westerly winds and influence the weather conditions in the middle latitudes.

Origin

There are various schools of thought about the origin of tropical cyclones. Fitzroy in 1863 postulated that tempeate cyclones originated because of the convergence of two opposing air masses of contrasting physical properties i.e. temperature, pressure, density and humidity. In 1911 Mr. Shaw and Lempfert gave "dynamic hypothesis" that temperate cyclones originated due to inflow of winds from all directions towards the centre. Latter on another theory known as "polar front theory" was put forth about the origin of temperate cyclones by Norwegian meteorogists V. Bjerknes and Bjerknes in 1918 based on the formation of fronts. When two contrasting and opposing masses (warm and cold) collide against each other and try to cross the domine of one another, the sustainable waves are formed which help in the origin of temperate cyclones. This theory holds that there are two currents of air having different temperatures and velocities. They flow parallel to each other and their surface of discontinuity is known as polar front.

Shape and Size

A temperate cyclone on an average is about three to four hundred miles round and about five to seven miles in height. They are generally V-Shaped depression. They are of different shapes and sizes e.g. circular, semi-circular, elliptical, elongated and V-shaped. All temperate cyclones are characterized by low pressure in their centre. The average diameter of a tropical cyclone is about 1900 km. Sometimes they are so large and extensive that they cover an area of 1000,000 km. The vertical extent of an average tropical cyclone is about 10 to 12 km.

VELOCITY AND SPEED

The temperate cyclones move eastward under the influence of westerly winds with an average velocity of 32 km per hour in summers and 48 km per hours in winters. Within the same cyclone, the force of wind is different in different portions of the cyclones. The winds are the strongest in the southern and eastern sections where their direction is identical with that of cyclone. The winds move parallel to the isobars upto a height of 5 km. It is a natural rule that slower the movement of the cyclone, greater is the stay of winds and heavier is the rainfall. The velocity of a temperate cyclone varies from place to place. It depends upon the season, location and varies from cyclone to cyclone. These are fastest in winter months and slowest during the summer season. They are faster in America than in Europe. Normally these cyclones travel with the prevailing winds i.e. westerliers and their usual direction is from west to east. But in the northern hemisphere they dip to the south over the continents while they are deflected towards the north over the oceans. These depressions may move in any direction but they always tend a little towards east.

WEATHER ASSOCITED

Due to the different types of air masses and varying temperature conditions different parts of temperate cyclone are associated with varying weather conditions. It experiences different weather conditions at the time of arrival and passage of warm front, warm sector, cold front and cold sector.

When the temperte cyclone arrives, wind velocity slows down considerably and air pressure decreases. Temperature increases with the coming of cyclone to a closer point. Wind direction changes from easterly to south easterly, the clouds become low, thick and dark.

Clouds become very thick and dark with the arrival of warm front of cyclone and heavy down pour starts. Since the warm air rises slowly along the front and hence the precipitation is gradual but for a long duration depending upon rising warm air. If the air is full of moisture there is sufficient precipitation.

The warm sector comes over after the passage of warm front and there is sudden change in the existing weather conditions. The wind acquires a southerly direction and the sky becomes clear. There is sudden increase in temperature and humidity but air pressure decreases, which bring occasional drizzles with pleasant and clear weather.

With the arrival of cold front the temperature shows a marked decrease increasing cold considerably. The cold air uses the warm air upward which changes the wind direction from south to southwest and west. Sky again becomes dark with cumulonimbus clouds which yield heavy rainfall, with thunder and lightening for a short duration.

Weather again changes remarkably with the passage of cold front and arrival of cold sector. The sky becomes clear and clouds disappear. The humidity and temperature decreases abruptly with the increase in air pressure. Wind direction changes from 45 degree to 180 degree and it becomes true westerly.

Occlusion

The frontal cyclones can not survive for a longer period as the western cold air displaces the warm air continuously. The result is that the warm sector gets reduced in size. With the passage of time the cold front becomes identified with the warm front and the cyclone dies. To this identification of cold and warm fronts we call occlusion or occluded front.

Generally, near the occluded front there is a good downpour due to the mixing of cold and warm air. There is definitely some difference of temperature between the two sides of this front. When a dense cold air mass moves into warm air sector, a cold type occlusion results. It lifts both the warm and cool air that underlies the warm air. On the head of warm air mass is the cold mass of air and the cool air approaches it. Due to its less dense nature it lifts the warm air but rests it upon the cold front. Hence, the warm air is left in a trough of cold and cool air masses. The cool air pushes under the warm air and overides the denser cold air.

Types

The temperate cyclones are generally divided into three principal divisions as under:

(1) Dynamic cyclones are the real temperate cyclones formed by the convergence of cold polar and warm, moist tropical air masses. They affect the weather conditions of areas of middle latitudes. They are dynamically produced e.g. convergence and inversion of two contrasting air masses into the territories of one another.

(2) Thermal cyclones are formed due to the development of low pressure centers on the continents in summer in the temperate regions. Hence, the winds blow towards the low pressure centres from all directions. Thermally induced temperate cyclones remain stationary at their places of origin, thus, different fronts are not developed. Hamphreys postulated that thermal or insolation cyclones are produced due to development of low pressure centres over warm water surfaces of seas surrounded by cold land surfaces during winters.

(3) Those cyclones which develop due to passage of cold winds over warm sea after the occlusion of main cyclones are known as secondary cyclones.

GENERAL CHARACTERISTICS

The period of a temperate cyclone form its inception to its termination is called the life cycle of cyclone which is completed through six successive stages.

(1) The first stage involves the convergence of two air masses of contrasting physical properties and directions. The air masses (cold and warm) move parallel to each other in its first stage and a stationary front is formed.

(2) In the second stage known as incipient stage the warm and cold air masses penetrate into the territories of each other. In this way a wave like front is formed.

(3) In the mature stage, the cyclone is fully developed and the isobars are circular.

(4) Due to the approach of cold front, warm sector is narrowed in extent, as cold front comes nearer to the warm front.

(5) This stage starts with the occlusion of cyclone when the advancing cold front finally overtakes the warm front and an occluded front is formed.

(6) This is the last stage in which warm sector completely disappears and the cyclone vanishes and dies out.

15
Oceanography

INTRODUCTION

The word ocean has been derived from the Latin word 'Oceanus' and Greek 'Okeanos'. The oceans are the major bodies of water and the seas are the minor ones which are partially surrounded by land. The oceans, comprising about 71 percent or 140 million sq. miles of the earth's surface, have tremendous potential. Besides being a great source of salts and foodstuffs, the tides can be harnessed to generate electric power.

The term "Ocean" also refers to any of the four major subdivisions of the earth, in order of size, they are Pacific, Atlantic, Indian and Arctic Oceans. The Pacific Ocean contains about half of the total ocean area and is larger in area than all the world's land combined. The oceans account for 97 percent of the earth's water area.

The equator divides the Atlantic and Pacific Oceans into North and South Atlantic and North and South Pacific. Occasionally, but incorrectly, the southern most part of the Atlantic, Pacific and Indian Ocean is called the Antarctic ocean. Sea, is a term commonly used to designate (i) the ocean (ii) a subdivision of ocean or (iii) a salt lake lacking an outlet to the Ocean (for example, the Dead and the Caspian Sea).

For many centuries the ocean has served as a source of food and as a medium of transportation. At present the ocean and its floor have become increasingly important sources of petroleum and natural gas, sulfur, salt, magnesium and other minerals. Food from the sea consists mainly of fish and other marine creatures, although certain seaweeds are widely used for animal feed and fertilizer.[61]

The first successful world wide deep sea expedition began with British expedition of the Challenger (1873-76). In the current century this subject has become so important that researches have been conducted by various institutions and international organizations.

Oceans of the World With Depth and Area

S.No.	*Oceans*	*Area in sq. miles*	*Depth in fathoms*
1.	Pacific	63,986,000	36.198
2.	Atlantic	31,530,000	30.143
3.	India	28,365,000	24.436
4.	Antarctic	—	—
5.	Arctic	5,541,000	18.456

The study of the oceans, including the nature of the water, its movements, temperature, the depth, configuration of its floor, the fauna and flora is called oceanography. The science of study of waters is known as 'hydrology' or Hydrography. Hydrosphere is that part of the earth forming the oceans, seas, lakes and ice sheets, covering the three quarters of the earth's surface. It is distinct from the atmosphere and lithosphere.

The oceans serve as store-houses of heat and water for mankind. They are the seat of incessant activities, the effects of which are extended to the modification of the land (near seas). Both the oceans and atmosphere influence each other, the former modifies the temperature of the air and the later carries water vapous with it. The depth of the oceans varies from place to place and ocean to ocean, which is measured in fathoms (6 feet). The older echo-sounding techniques have now been replaced by radar sounding and electrical echo-devices to find the precise depth of the oceans and relief.

RELIEF OF THE OCEAN FLOOR

The ocean basins are in many ways similar to the land surface. These are submarine ridges, plateaus, canyons, plains and trenches. According to the depth, an ocean can be classified into four zones or sections viz. (1) continental shelf of varying width fringing the continents, (2) continental slope descending to the ocean basins, (3) broad expanses of plains smoothed by deposites, and (4) depressions usually linear, known as ocean deeps. These are illustrated in detail one by one as under:

(1) Continental Shelf

It is the sea bed that border the continents, usually covered by shallow water, having an average depth of 100 fathoms. It takes the form of a shelf or ledge sloping gently downwards from the coast. At the edge it change in depth from about 65 to 200 fathoms. It also varies in width from 20 to 100 miles. It is few miles in the north Pacific Ocean to over 100 miles off N.W. Europe. On the contrary extremely mountainous coasts have about no continental shelf i.e. very narrow, as there is a quick transition from high land to deep water. The Rocky and Andes mountains (Americas) are the classic examples. Where it is widest, the angle of slope is generally least (less than 1 degree).

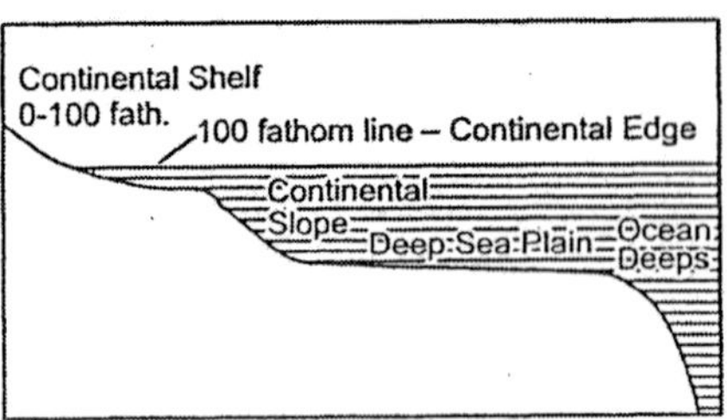

Fig. 1. Relief of Ocean Floor.

The continental shelves are basically the extended form of continental platforms originated by marine erosion (waves and currents) of continental margins. They are also the result of fluvial deposits as prolonged deposition and sedimentation construct most extensive continental shelves under calm waters of the sea. Many oceanographers propound that the continental shelves are formed due to the compressive forces which causes subsidence of continental margins. Sometimes parallel faults are created in the continental margins, which also causes subsidence of these land areas. Hence, the continental shelves formed are of tectonic origin. Besides, the extensive continental shelves in the North Sea and near Newfoundland are due to deposition of debris that followed the melting of ice sheets of the great Ice Age.

(2) The Continental Slope

It is a marked slope from the edge of the continental shelf to the deep sea or abyssal plain. It has a varied slope of 200 to 2000 fathoms, in some places much further e.g. to 5000 fathoms off the Phillipines. The angle of slope is generally between 2 to 5 degrees but sometimes it varies from 5 degree to 60 degrees e.g. 40 degree near St. Helena, 30 degree Spanish coast 62 degree near St. Paul and 5 to 15 degree near Calcutta coast India. In other words it is a transition zone between the continental shelf and the ocean trough and is comparatively shorter in extent.

Through the continental slope one finds the sub-marine canyons or channels, continuing sometimes across the continental slope. Some of these canyons seem to be extensions of river courses of the present day. They were produced when the sea level was considerably lower than it is now. Canyon at the mouth of Hudson river (USA) is a classic example which extends about 320 km into the sea. These trenches are generally transverse to the continental shelves and the coasts.

The origin of continental slopes has been related by various authorities to erosional, tectonic and aggradational processes. The erosion theory of the origin of continental slope is based in the presence of submaring canyons This theory states that the continental slopes are the results of marine erosion chiefly by waves. In some cases the continental slopes have tectonic origin too, as these slops are formed by faulting and folding. Some geologist are of the belief that these slopes are formed due to bending and warping of continental shelves. While the others take it the result of fluvial deposition and sedimentation.

(3) The Deep Sea Plain

The deep sea or abyssal plains or ocean troughs are the

extensive and fairly level areas about 4,000 to 6,000 meters under the surface, that forms the majority of the ocean floor. At some places they descend to the ocean deeps. These plains cover about 76 percent of the total area of the ocean basins but it varies from 55 percent (Atlantic Ocean) to 80 percent (Pacific and Indian) in different ocean. These plains have extensive submarine plateaus, ridges, trenches, basins and ocean islands. These deep sea plains often have narrow and elongated ridges as its significant features, known as submarine ridges. They have steep side slopes which at times reach the sea level and even project above water surface and appear as islands. Mid-Atlantic ridge, East Pacific ridge and Mid-Indian ridge are the classic examples.

(4) Ocean Deeps

Ocean deeps are the deepest trenches and depressions in the ocean troughs, generally located parallel to the coasts facing mountains and along the islands. Ocean deeps are sub-divided into two categories viz. (1) 'deep' which are deepest zones but less extensive, (2) long and narrow linear depressions called 'trenches', both of these having steep slopes. They vary in size and number from ocean to ocean. The Pacific Ocean has 32 deeps, the largest number while the Atlantic and Indian Ocean has 19 and 6 deeps respectively. Marina trench in N. Pacific Ocean near Phillipines is the deepest about 11 km (more than 36000 feet) in all the ocean deeps. Thus it is obvious that the ocean trenches are greater in magnitude than the highest mountains on land, for the highest peak Mt. Everest is only 29,028 feet high a.s.l. The other significant deeps of Pacific Ocean are Mindanao deep 35000 ft, the Tonga trench 31000 and Japanese trench 28000 ft.

As a matter of fact, the ocean basin floor is not of uniform depth throughout, neither it is continuous. Its continuity is interrupted by deeps, trenches and islands. Thus the relief of the ocean floor is almost similar to that of the land surface or lithosphere viz. mountains, valleys canyons, plateaus and plains.

TEMPERATURE OF THE OCEAN WATER

The oceanic water temperature depends mainly on the sun, as it is the principal source of heat and energy. The radiant energy transmitted from the sun in the form of electromagnetic short waves is known as 'radiation'. This radiated heat of the sun is received by the surface of the earth and ocean surface is called 'insolation', which is responsible for heating the oceanic water. But the process of heating and cooling of water is quite different from that of land.

CAUSES OF TEMPERATURE VARIATION

The specific heat of water is five times that of land. The evaporation is a cooling process for water which is a continuous process. Due to these reasons, water takes longer time to be heated and cooled in comparison with land. That is why the ocean waters as well as the underground waters in winter are warmer than the surrounding land areas and in summers, it is cooler. The temperature of the ocean water varies from ocean to ocean and from place to place within the same ocean. The factors responsible for this variation are as follows:

(a) Latitude

Due to the small insolation, short summers and slanting rays of the sun the temperature of polar latitudes is seldom more than 29 degree F. Whereas in tropical regions it is as high as 95 degree F. In other words the temperature of surface waters decreases from equator towards the poles.

(b) Prevailing winds

The off-shore winds blowing from the land towards the sea drive warm surface waters resulting into upwelling of cold water from below. This replacement of ocean water results in longitudinal variation of temperature. Contrary to this, the onshore winds blowing from sea towards land pile up warm water near the coast and thus raise the temperature.

(c) Depth

The greater the depth of the oceans more is the decrease in temperature of the water is a natural rule. As matter of fact, the sun rays can penetrate only upto 617 fathoms. Hence, the temperature of the water below this depth abruptly falls.

(d) Uneven Distribution of Land and Water

The oceanic water temperature varies from northern to southern hemisphere because of the dominance of land in former and water in the later. The temperature of surface water of the oceans in the north is higher than in the southern oceans. The temperature in the enclosed seas in low latitudes becomes higher because of the influence of surrounding land areas than the open seas. The typical example being the Red Sea with 37.8 degree C and the Persian Gulf with a temperature of 34.4 degree C both located in equatorial regions where the average temperature of surface water is 26.7 degree C.

(e) Salinity

The temperature of the ocean water varies with the variation of the salinity of the ocean water. The greater the salinity the higher is the temperature is a natural rule. Thus it is highest in the equatorial region and decreases towards poles. It is also comparatively higher in land-locked seas than open oceans.

(f) Ocean Currents

The temperature of the ocean water is greatly influenced by the currents. Warm currents raise the temperature of the regions they visit, whereas the cooler currents lower down the temperature. Generally warm currents originate in equatorial belt and flow towards the poller belt as surface current. On the contrary the colder current flows as under current towards the equator from pole. Hence, they influence the temperature of the affected areas. The classic example being the Gulf Stream (warm) which raises the temperature of eastern coasts of N. America and western coasts of Europe. Besides Kuro Sivo (warm) raises the temperature of Alaska, a cold region.

(g) Change in Seasons

The temperature of the ocean water also varies according to the change of seasons. The duration of the days and nights is very much affected with the change in seasons. Thus more the duration of the days the more will be the insolation and more the temperature of the oceanic water and vice versa.

DISTRIBUTION OF OCEAN WATER TEMPERATURE

The distribution of temperature of ocean water is studied in two ways viz. (1) horizontal distribution of surface waters and (2) vertical distribution from surface water to the bottom.

The annual variation of the surface temperature depends upon different factors such as annual variation in radiation, ocean currents, and the prevailing winds. In equatorial and polar regions the annual variation remains very small. On the other hand in 30 degree to 40 degree latitudes the radiation is maximum.

Horizontal Distribution

The average temperature of the oceanic surface water in about 27 degree C. It gradually decreases from equator towards the poles at the rate of 0.5 degree F per latitude. The average annual temperature of northern oceans is 19.4 degree C and of southern ocean is 16 degree C. This variation in the northern and southern hemisphere is due to the unequal distribution of land and water.

Due to the warm ocean currents the decrease of temperature with increasing latitudes is very low in southern Atlantic ocean as compared to northern Atlantic ocean. According to the Krumel, the highest temperature of oceanic surface water is in 5 degree to north and south of the equator whereas the lowest temperature is found in 80 degree to 90 degree north and south pole.

The difference between air and sea surface temperature causes fogs over the seas and the oceans. This happens when warm air passes over a cold sea surface having the temperature below dew point.

Vertical Distribution

The temperature of the ocean waters also varies vertically with increasing depth. It decreases rapidly for the first 200 fathoms at the rate of 1 degree F for every 10 fathoms. After 500 fathoms the drop of temperature is very less 1 degree F for every 100 fathoms. In the ocean deeps below 2000 fathoms the water is uniformly cold but above freezing point.

It may be pointed out that maximum temperature of the oceans is always at their surface because it directly receives the insolation. The heat is transmitted to the lower sections of the oceans through the mechanism of conduction. Infact the solar rays very effectively penetrate upto 20 meters depth and they seldom go beyond 200 meters depth. Thus, on this basis the oceans are vertically divided into two zones. (1) Photic zone, which receives the solar rays and (2) Aphotic zone, which extends from 200 meters to the bottom and never receive any solar rays.

Salient Features of Distribution of Temperature

(1) The rate of decrease of temperature is not uniform, though it decreases with depth. The change in sea temperature is negligible below the depth of 200 meters.

(2) The diurnal range of temperature cease after the depth of 5 fathoms whereas the annual range cease at 100 fathoms.

(3) The rate of decrease of temperature from equator towards the poles is not uniform. The temperature at the ocean bottoms is uniform from equator towards the poles.

(4) The enclosed seas of high latitudes register inversion of temperature i.e. the temperature of sea surface is lower than the temperature below.

SALINITY OF OCEAN WATER

The degree of saltness of the oceans and seas, lakes and rivers, usually expressed as the number of parts per thousand or the weight of salt dissolved in 1000 parts of water is known as salinity. The mean salinity of sea water is 35 per 1000. It varies from over 40 per 1000 in the Red Sea to about 30 per thousand in the polar seas. Owing to the heavy rains the equatorial region has relatively less salinity. Bordering this belt are the high salinity regions as the dry Trade Winds cause rapid evaporation, thus, turn the surface waters into salt. Beyond this belt the salinity again decreases with the increase of precipitation and less evaporation. The salinity is measured by the latest device known as Electric Salinity Meter.

Although river contains a small quantity of mineral salts in solution, it is comparatively fresh. The Baltic Sea has only 12 and Black Sea 17 per thousand salinity because they receive many large fresh water rivers and heavy rains. On the contrary the Red Sea has 40 and Mediterranean Sea 37.35 per thousand of salinity because they receive no large revivers, less rainfall and have great evaporation. The seas and lakes with no outlet to the oceans have a still higher salinity. The salinity of Dead Sea (Israel) is about 250 and Lake Van, in Asia Minor 330 per thousand as salt in these basins has accumulated since their formation. Of the 35 parts of dissolved salt contained on the average in 1000 parts of sea water, over 27 parts consist of sodium chloride (common salt). The other constituents being magnesium chloride, magnesium sulphate and calcium sulphate.

The salinity of the water not only effects the density but also its colour and temperature. If the salinity is high, the colour of the water will be deep blue, but if it is less, the colour will be green. The temperature of the water increases with the increase of salinity. Hence the water with lower mineral contents has lower temperature.[54]

COMPOSITION

As the sea water is an active solvent, it contains a complex solution of several mineral substances in dilute form. The salt is gradually increasing because it is brought from the land every year. The estimates of Joly, Murray and Clark put the total salt in oceans and seas at 50 billion tonnes, 5 billion tonnes and 2.7 billion tonnes respectively. According to Joly if all the salts of all the oceans are dried up and are spread over the whole land, this will form 152.4 meters thick layer. If all the salts are removed from the oceans and seas, there will be a fall in sea level by 30.48 meters (100 ft). During the Challenger Expedition in 1884, Dittmar reported the existence of 47 types of salts in sea water. The seven of them are most important, the table elaborates as follows:

Table Salts in the Oceans

S.No.	*Salt*	*Symbol*	*Amount (per thousand)*
1.	Sodium Chloride	Na cl	27.2
2.	Magnesium Chloride	Mg cl	3.8
3.	Magnesium Sulphate	$Mg\ So_4$	1.6
4.	Calcium Sulphate	$Ca\ So_4$	1.2
5.	Potassium Sulphate	K_2So_4	0.86
6.	Calcium Carbonate	$Ca\ Co_3$	0.12
7.	Magnesium Bromide	$Mg\ Br_2$	0.07

About 68 percent of the salt in sea water is the sodium chloride or common salt. There are lesser amounts of magnesium chloride 14 percent, sodium sulphate 11 percent, calcium chloride 3 percent and other salts. The freezing point of saline sea water is 28.6 degree F as compared to 32 degree F for fresh water. The average specific gravity of sea water is 1.045. This is 4.5 percent greater than that of fresh water. The specific gravity varies somewhat with salinity, temperature and depth. This greater density makes it easier to float and swim in sea water.[69]

SOURCES

The land is the chief source of salinity which is washed away by rain water and rivers and finally deposited in the oceans. Volcanic ashes also provide some salt to the oceans. The mineral salts come from eroded rocks. Much of which is ultimately carried to the sea by rivers, glaciers and winds and thus several billion tonnes of salt are added to the sea each year.

The two hypothesis have been put forward for the origin of salinity of the oceans. According to the first hypothesis the salt was present in the water vapours before the cooling of the earth. When the earth cooled down the salt was added to the ocean waters. According to the second hypothesis the oceans became gradually saline by the earth's crust, as it was washed by rain waters and the rivers transported the dissolved minerals to the sea.

CONTROLLING FACTORS

There is a wide range of variation in the oceanic salinity. The controlling factor affecting the amount of salt in different oceans are the evaporation, precipitation, river water, prevailing winds, ocean currents and waves.

(a) Evaporation

The greater the evaporation, the greater is the salinity of the ocean water and vice versa. Evaporation due to high temperature with low humidity causes more concentration of salt. Highest salinity is thus found in tropics of cancer and capricorn.

(b) Precipitation

The higher the precipitation, the lower is the salinity and vice versa is the rule of the nature. This is why equatorial regions with highest rainfall have the lowest salinity. The polar regions also have lowest rate of salinity because of presence of snow and ice.

(c) River Water

The fresh water of rivers have very less salinity. That is why the Baltic sea has only 17 per thousand salinity. Salinity decreases with maximum run-off during rainy season and increases in the season of minimum run-off.

(d) Prevailing Winds

The winds are also one of the significant factors which tend to increase or decrease the salinity of the oceans. Anticyclonic conditions with stable air and high temperature are favourable for the increase of oceanic salinity. Such conditions prevail in sub-tropical high pressure belts. The winds drive away saline water to less saline areas, thus results in redistribution of salinity in ocean. This is why the Gulf of Mexico has a high salinity of 37 per thousand.

(e) Currents and Waves

The ocean currents play a vital role in redistribution of salinity, the classic example of which is the high salinity of Gulf of Mexico. The equatorial warm currents drive away salts from the western coastal areas of the continents and accumulate them along eastern coasts. The extension of Gulf Stream increases salinity along north western coasts of Europe. On the other hand the salinity is reduced along the N En. Coasts of N. America due to cool Labrador Current.

HORIZONTAL DISTRIBUTION OF SALINITY

By Horizontal distribution we mean the variation of salinity on the surface of the ocean water. In other words, the distribution of oceanic salinity in relation to latitudes or regions.

(1) Latitudinal Distribution

Though the salinity decreases from equator towards the poles, yet on the basis of latitudinal distribution of salinity four zones have been identified e.g. (i) equatorial zone of relatively low salinity due to excessive rainfall, (ii) tropical zone (between 20 degree to 30 degree N latitude) of maximum salinity due to low rainfall and high evaporation, (iii) temperate zone of low salinity, and (iv) polar zone of minimum salinity.

(2) Regional Distribution

Jenkins has divided the oceans on the basis of salinity variations into three categories e.g.. (i) seas having salinity above normal (Red Sea 40 per 1000, Persian Gulf 37.38 per 1000 and Mediterranean sea 37.35, Gulf of Mexico 37 per 1000), (ii) seas having normal salinity (Caribean Sea 36 per 1000, Bass Strait and gulf of California 35.5 per thousand each) and (iii) seas having salinity below normal (Arctic Ocean 20 per 1000, Bering Sea 28 per 1000, China Sea 25 per 1000, Japan Sea 30 per 1000 and Baltic Sea 12 per 1000).

VERTICAL DISTRIBUTION

Vertical distribution of salinity is the variation of salinity from the surface of the ocean water towards the bottom of the ocean. No definite trend of distribution of salinity with the depth can be spelt out. As for instance the surface salinity at the southern boundary of Atlantic is 33 per 1000 but it increases to 34.5 per 1000 at a depth of 200 fathoms. At a depth of 600 fathoms it reaches to 34.75 per thousand. On the contrary, the surface salinity is 37 per 1000 at 20 degree latitude but it decreases towards the depth.[55]

CIRCULATION OF OCEANIC WATERS

The ocean water though seems still is never so, but as a matter of fact it is always in motion. It is disturbed by certain factors which give it a movement. These factors or forces try to disturb its level and equilibrium and the nature tries to restore its level again. It results in the movement of the oceanic water of various types possessing different characteristics. Some of these movements are visible such as waves, tides and currents and some are obscure like drift and creep. The visible movements (waves, tides and currents) are of significant importance among the other movements of the ocean. The tides are, in fact, the most important of all, as it affect the whole water mass from the surface to the bottom of the sea.

WAVES

Waves are the vibrations or the swinging of surface water of the ocean, backward and forward. It is an oscillation of water particles on the surface of the ocean produced by the friction of the winds upon the water surface. The crests and troughs are formed by rise and fall of the surface water of the ocean, when winds blow over it.

The size and strength of the wave is determined by the speed and duration of the wind that causes it. Under a strong wind storm, the waves may be a hundreds of feet in length and a few tens of feet in height. The waves move to and fro and do not progress horizontally. The waves dash against the shores as 'breakers' when they approach the shores with retarted speed.

All the waves are forced by the wind energy, hence, they are known as 'forced waves'. When these are of large size and strength, they are called as storm winds and if the waves are away from the place of disturbance, they are known as 'ground swell' or free waves. The waves, as a rule of nature, return back to the same point from where it has taken it birth, thus, are known as 'closed waves'. But sometimes, in an open ocean the wind induces as forward movement in the waves which is known as 'drift' or 'open waves'.

The nature of the wave is defined by the following salient features of a wave.

(a) Height

The distance from crest to trough, that may be as much as 40 to 50 feet is the height of a wave.

(b) Length

The distance between two successive crests is called as the length of a wave. The longest measured wave was 3700 feet.

(c) Period

The time taken for the wave-form to move the distance of one wave length.

(d) Velocity

The speed of the forward movement of an individual crest is known as the velocity of the wave.

(e) Steepness

The ratio of its height to its length, that increases as the wave enters shallow water until it reaches 1:7 when it breaks.

(f) Energy

The energy of a wave depends directly upon all the preceding features including fetch. The average pressure exerted by a large wave in winter on an exposed coast is nearly a ton per sq. ft. (western Ireland). During the storm it may be more than three times as great.

The waves are the significant erosive agents and their role in marine erosion is the most dominant. The erosive action of a wave is performed through hydraulic action, corrosion, attrition and solution processes. The waves are destructive for shippings as the crafts suffer serious damage when high waves rise in the sea. But they can prove inexhaustible source of potential energy if exploited and harnessed in a proper way.

TIDES

The periodic or alternate rise and fall of sea water due to the gravitational forces of the sun and moon are known as tides. The rising of waters is known as the 'high tide' while the falling of waters is called the 'low tide' or 'ebb tide'. It has been observed that the interval between two high tides and between two low tides is about twelve hours.

It is a matter of fact that the earth rotates from west to east and revolves around the sun. Similarly the moon rotates and revolves around the earth. Both the earth and moon follow an elliptical orbit so that the distance between the moon and the earth and the earth and sun varies from time to time and month to month. It is evident that the gravitational pull of the moon or the sun will be maximum when the earth's surface is nearer to them and minimum when the earth is farther from them. Consequently, the water of the sea facing to the moon or sun is pulled and high tide occurs and the low tide is formed at the opposite side of the earth.

The role of the moon in causing tides is most significant, though it is smaller than the sun, it is much nearer the earth. The gravitational pull of the moon is two and a one fourth times more than that of the sun. The pull of the moon is most felt on water as it is loose and less on land which is of compact nature. This accounts two high tides, one direct in front of moon and the other indirect on its opposite side.

The earth rotates on its axis on account of which each portion of the earth is at one time facing the moon and at another the opposite side. Thus, in 24 hours each place of the earth has two tides. The interval between them is 26 minutes more than 12 hours due to one complete revolution of moon around the earth in 28 days. When the sun, the moon and the earth are in the same line their gravitational forces work together and high tides are formed on the contrary, the low tides come into existence when the sun and moon make a right angle with the earth and both of them exert the force against each other.

(1) Spring Tide

It is the tide of great amplitude which is caused when the sun, the moon and the earth are almost in a straight line. This position of the three heavenly bodies is called as 'syzygy', responsible for very high tide. When the sun and the moon are in one side of the earth in a straight line, the position is called as 'conjuction' (solar eclipse). On the other hand the position is known as 'opposition', when the earth comes in between the sun and moon. The former position takes place during new moon while the latter position is experienced in full moon. The gravitational pull of the sun acts in the same direction as that of the moon and therefore reinforces it giving it 20% more height than the normal tide. Such tides occur twice in each month at new and full moon at fixed timing.

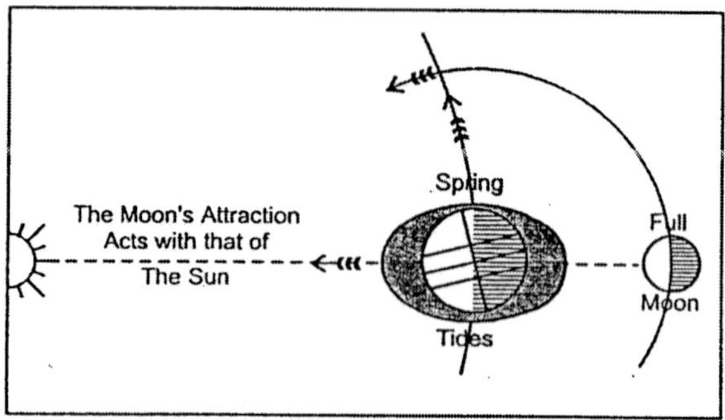

Fig. 2. Spring Tides on Full Moon.

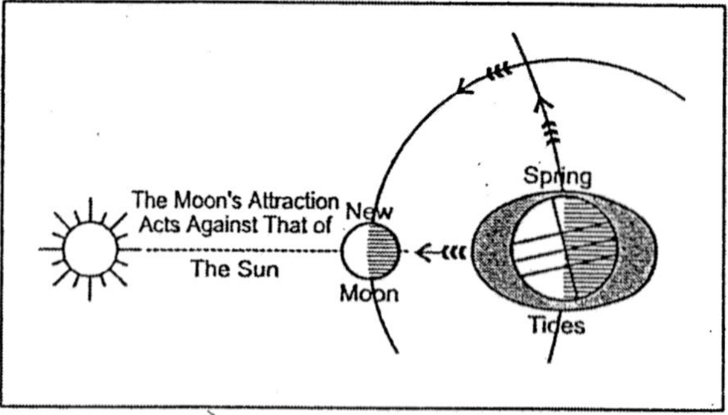

Fig. 3. Spring Tides on New Moon.

(2) Neap Tide

These are the tides of small amplitude formed when the gravitational pull of the sun opposes to that of the moon. The sun, the earth and the moon come in the position of 'quadrature' i.e. a right angle on seventh or eighth day of every fortnight or a month. Thus the tide producing forces of the sun and moon work in opposite direction to cause a low tide known as neap tide. These tides are about 20% lower in height than the normal tides and occur during the first and last quarter of the moon.

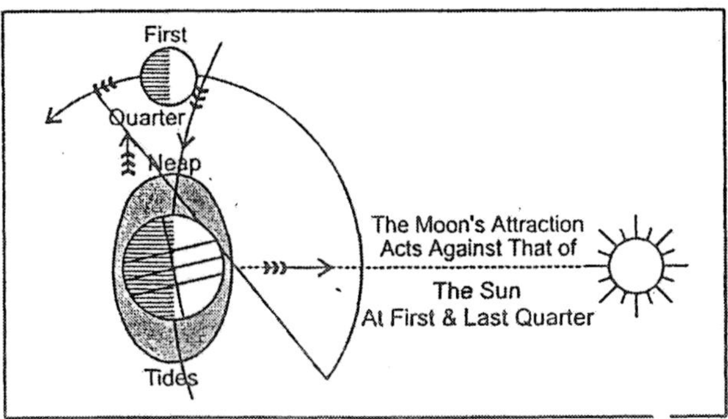

Fig. 4. Neep Tides.

THEORIES OF TIDES

There are various theories about the origin of ocean tides put forth by different geographers from time to time. These hypothesis include progressive wave theory by William Whewell (1833), stationary wave theory by R.A. Harris, equilibrium theory by Issac Newton (1687), Dynamic theory by Laplace (1755) and Canal theory by G.B. Airy (1842).

(1) Progressive Wave Theory

Willium Whewell propounded the progressive wave theory in the year 1883 about the origin of tides. This theory holds that two tidal waves are produced in the southern ocean where there is no interruption of land for about 119 degree of longitude. Hence, the attraction of moon is greatest there. These waves travel with the moon till they reach the cape of Good Hope where a secondary wave is formed. This secondary wave runs along the shores of the Atlantic Ocean.

The tidal waves are generated in the southern ocean under the influence of tide producing force of moon. These waves are called primary waves which move from east to west. When the westward movement of primary waves is obstructed by the land masses, the waves generated are called as secondary waves. It may be pointed out that the tidal waves after being originated in the southern ocean progressively move northward with continuous lag of time and dissipation of wave energy. The time of tides is progressively delayed northward along the longitude. This is why there is difference of time of tide at different places on the same longitude.

(2) Stationary Wave Theory

Mr. R.A. Harris propounded the stationary wave theory as opposed to the progressive wave theory. According to him the tides occur due to stationary waves which originate independently in each ocean or it is a regional phenomena. His theory works on the analogy of water in a tank or tray. If the tank or tray is tilted or rocked from one side to the other, the water begins to swash round the middle axis, of the front. In other words, it generates oscillation in the water. Such oscillations are called stationary waves.

Based on the above theory, the tidal forces of the sun and moon cause oscillation in the ocean waters around a central point because of the rotation of earth. This oscillating systems are nothing but stationary wave movements and they correspond to the periods of pull of the moon that produces tides. This system is affected by the depth, configuration and length of the ocean basins and the spread of the rotation of the earth. Greater the depth of the ocean, higher the stationary waves are generated.

CURRENTS

The currents are the general movements of water within the ocean which flow regularly, constantly and in a definite direction. Their movement is clearly visible over the otherwise calm surface of ocean water. The ocean currents are very important and most powerful as they drive oceanic waters for thousands of kilometers away from their place of origin. Currents are of two types e.g. warm and cold, but on the basis of velocity, dimension and direction they are called as drifts, currents and streams respectively. The gushing streams of waters have a greater width and depth and move with greater velocity e.g. Gulf stream.

CAUSES

The ocean currents are the result of combined effects of various internal as well as external factors. These factors include the prevailing winds, gravitational force, rotation of the earth, temperature variations of water, pressure gradient, salinity variations, relief of ocean floor and seasonal variations etc. Some of them are discussed below.

(a) Temperature Difference

The higher the temperature, the lesser will be the density of ocean water is a natural rule. Thus, due to the high temperature in the equatorial region the density of water decreases because of expansion. On the contrary, the density of the polar waters is comparatively greater because of intense cold. Consequently, the warm surface currents move from equator towards the poles and reverse sub-surface currents move from poles to equatorial region, making a complete circulatory system of ocean water.

(b) Salinity Differences

In the land-locked seas density follows salinity as the water is more saline due to greater evaporation and lesser receipt of fresh water. Hence, this higher density results in an undercurrent of salt water from enclosed sea towards the open ocean. On the contrary, the surface current of less saline water moves from ocean to the enclosed sea to fill up the gap. Such currents flow from Atlantic Ocean to Mediterranean sea via Gibralter Strait as surface currents and vice versa as dense undercurrent. Similarly, the surface and sub-surface currents are generated between the Red sea and the Arabian sea via Babel Mendab Strait.

Besides the above factors prevalent winds drive the ocean waters in a continuous flow in front of them. The currents follow the main direction of the planetary wind systems. In lower latitudes the ocean currents accompany the trade winds towards the west. The Indian Ocean currents follow the S.W. and N.W. monsoon winds as they change their course according to the season. The rotation of the earth and its configuration also affect the direction of currents. According to the well known Ferell's Law the direction of currents is clockwise in Northers Hemisphere and anticlockwise in Southern Hemisphere. The east west bulging land masses deflect the currents towards the north and south in the equatorial region is the clear affect of the configuration of land masses in diverting the currents.

CURRENTS OF THE ATLANTIC OCEAN

In the centre of the Atlantic Ocean near equatorial region the two most important currents are the North Equatorial Current and South Equatorial Current. The N. Equatorial Currents is formed between the equator and 10 degree N. latitude. It flows from east to west but being obstructed by the land barrier of the east coast of Brazil, is bifurcated into two branches e.g. (a) Bahamas current and (b) Carribean current. The Bahamas current also known as Antilles current is diverted northward and flows to the east of West Indies. The Carribean branch enters the Gulf of Mexico and latter becomes the Gulf Stream.

The South Equatorial Current flows from the western coast of Africa to the eastern coast of South America. This current is the continuation of the Benguela current and is bifurcated into two branches by the bulge of Brazil. The northward branch taking N.W. course merges with the N. Equatorial Current. On the other hand the second branch turns southward as Brazilian current under the stress of trade winds.

(a) Gulf Stream

The Gulf Stream a warm current originates in the Gulf of Mexico (20 degree north latitude) and moves in north easterly direction coast of Europe (70 degree north latitude). The famous river Mississippi meets the Carribean Ocean Current a branch of north equatorial current in the Gulf of Mexico. The accumulated water gushes forth from the straits of Florida, leaving the stream half a mile deep and thirty miles wide. It moves with a speed of 5 to 8 miles per hour. The water of the Gulf Stream has surface temperature of about 80 degree F. Its temperature and speed declines as it reaches in the middle of Atlantic Ocean where it is known as North Atlantic Drift.

The Gulf Stream is divided into two branches. One branch passes round the British Isles, washes the shores of Norway and ends in Arctic Ocean. On the other hand, the second branch moves eastward and near Spain it branches off into two portions. One flows towards south which is called as 'Carribean Current' and the other moves northward and joins N. Atlantic Drift. The Gulf Stream modifies the temperature of western Europe and British Isles.

(b) The Labrador Current

The Labrador current, a cold counterpart of Gulf Stream originates in the Baffin Bay and Davis Strait. It flows through the coastal waters of Newfoundland and Grand Bank and merges in the Gulf Stream (50 degree W longitude) producing a dense fog which is dangerous for shipping. Besides it brings with it a large number of big icebergs which present effective hindrances in the oceanic navigation, known as 'cold wall' (New York).

CURRENTS OF PACIFIC OCEAN

The currents of Pacific Ocean are almost similar to that of Atlantic Ocean. There are two important currents e.g. North Equatorial Current and SouthEquatorial Current. The North Equatorial Current takes its birth from West coast of Mexico and flows westwards and reaches the Phillipines after covering 7500 nautical miles. It turns to the north passes along the Japanese islands known as 'Kuro Siwo'. It modifies the temperature of Japanese islands. Its branch reaches the British Columbia as West Wind Drift. It washes California coasts. Its counterpart is the Cold Curile Current, which in its origin and effects is very much similar to that of Labrador Current of Atlantic Ocean.

In the Southern Pacific, the South Equatorial Current is originated due to the influence of south east trade winds and flows from east to west. It reaches the eastern coast of Australia where it is called as East Australian Current when moves southward. Its second branch moves along the northern coasts of Australia. The cold current along the west coast of South America is called Humboldt or Peru Current. Its velocity is about 15 nautical miles per day with a temperature of about 16 degree C.

CURRENTS OF THE INDIAN OCEAN

The currents of Indian Ocean are directly controlled by the monsoons and to some extent by the landmasses. The currents of northern Indian Ocean changes their course of flow twice a year due to north east and south east monsoons. During winter north east monsoon winds blow from land to the ocean. As such the westward blowing northeast monsoon currents are produced between Andaman and Somalia. These currents flow by the Indian coastline in a westerly direction known as N.E. Monsoon Drift. It takes a southward turn along E. African coast and then moves westward as Indian Counter Current (Cold).

In summer, the reverse is the direction of monsoon winds and they flow along the Indian coastlines taking an easterly course. Thus, the south west monsoon current is developed which moves along the southern coast of India and finally merges with the main equatorial current. The currents of southern Indian Ocean are not affected by any seasonal changes of monsoon winds. The Indian Equatorial current flows in a east-west direction from Australian coast to African coast between 10 degree to 15 degree south latitude. Its eastern arm along Australian coast is called as the West Australian Current, while its western arm branches off in three divisions. One branch moves northward as W.W. Monsoon Current and the other two move southward as the Mozambique Current passing by the coast of the mainland of east Africa and the Madagascar Current making a loop round the Madagascar island. Both these currents join again to form a single current known as Agulhas, which washes the shores of S.E. Africa,. It takes an easterly turn and merges into the Antarctic Drift.

EFFECTS OF OCEAN CURRENTS

The warm currents do not allow the temperatures of colder areas to fall but rather they make them relatively warmer in winter months. The ideal and favourable climate of western coasts of Europe is due to the effects of the north Atlantic warm current an extension of Gulf Stream. The temperatures of coastal countries like Great Britain, Norway, Sweden, Denmark and Netherlands are higher during winter as compared to the place of their respective latitudes. The Gulf Stream raises the temperature of Atlantic and coastal plains of USA during summer and is responsible for hazardous weather conditions which results in several deaths every year.

The warm currents maintain the temperature of the ocean water while the cold currents lower the temperature of the affected areas and cause snowfall. Labrador, Kurile and Falkland cold currents result in heavy snowfall in the affected areas during winters. On the contrary, the warm currents transport the warm waters of the equatorial and tropical regions to colder areas and hence balance the temperature of the polar zones.

The winds blowing over the warm currents pick up moisture and bring heavy rains in the coastal areas, the north Atlantic Drift a warm current bring sufficient rainfall along the western coast of Europe. The Kurosiwo another warm current of the Pacific Ocean brings heavy rainfall along the eastern coast of Japan. On the other hand, the Kalhari desert of S. Africa is the result of Benguela cold current and the Acatama desert of Wn. South America is the result of Peru cold current as they discourage the rainfall. The dense fog of Newfoundland is the result of meeting warm and cold currents which is very dangerous for navigation.

OCEAN DRIFT AND CREEP

Ocean drift is a slow movement of surface water at sea or on a lake caused by the wind. Like currents these are broad and shallow. Besides the winds they are also caused by the differences in temperature, pressure, salinity and density of ocean water, to which they try to equalize. Certain currents on becoming broader and shallower turn into drift such as N. Atlantic Drift and W. Wind Drift (Pacific). Antarctic Drift is the classic example of ocean drift.

Ocean creeps are the under-currents which flow by the sinking of cold and heavy water as its density increases. The water of the poler areas travels towards the equator in slow creeps below the surface water as an undercurrent to equalize the temperature and density of water.

MARINE DEPOSITES AND CORAL REEFS

Ocean Deposit

The rocks underlying the deep sea plain of the ocean floor are always covered with deposites known as muds, oozes and clays. These deposits consists largely of the remains of various organisms which exist in the surface waters and on dying settle to the bottom. The other sources of ocean deposites are rivers, winds, burning meterors etc. They also contain volcanic dust, which has been carried by the wind and has like wise sunk to the ocean bed. The oozes are classified as calcareous or siliceous, named according to the organisms abundant in them. The typical example of calcareous oozes are Petropod ooze and Globigerina ooze while of siliceous oozes are Radiolarian and Distom oozes.

The ocean deposits found at various depths of ocean floor and according to their origin are grouped into two main divisions as under:

(1) Terrigenous Deposits

All those marine deposits which are derived from the land and consist mostly of boulders, pebbles, sand, gravel and muds are known as terrigenous deposits. They are deposited in their order of fineness on the shores in the shallow continental shelf, continental slope and deep ocean floor. The coarser material of bigger size such as boulders, pebbles and shingle are deposited near the shores. The shallow water deposits consists of gravel, sand and silt, while the finest particles of clayey muds (blue, yellow and green) are deposited in the deep sea floor.

(2) Pelagic Deposites

The deposits which cover the bed of the pelagic zone of the ocean, formed largely of the remains of marine plants and animals, which in life float in the surface waters. These deposits cover about 75 percent of the oceanic area both organic and inorganic in nature. They are like shells in form and chalky in structure, having a predominance of line or silica.

(a) Oozes

Organic deposits in ocean basins are known as oozes, which consist of calcereous and siliceous in formation. An ooze is formed from the remains of animals (skeletons) and plants which float on sea surface. They are subjected to decomposition and chemical changes to mud and sands and are deposited on the sea floor.

Calcareous oozes are formed by the animal skeletons containing lime content in abundance. They are seldom found at greater depth because of their high degree of solubility. Petropod oozes contain 80 percent of calcium carbonate and is mostly found in tropical oceans between the depth of 300 to 1000 fathoms. They predominate in western and eastern parts of the Pacific Ocean, Mediterranean submarine ridge and Indian Ocean. Globigerina ooze formed from shells of variety of foraminifera and germs called Glbigerina. It dries up in milky white, blue, grey, yellow and green powder. The chemical composition reveals 64.46 percent of calcium, 1.64 percent of silica and 3.3 percent of minerals. It is mostly found between the depths of 2000 to 4000 fathoms in tropical and temperate zones of Atlantic Ocean, eastern and western zones of India Ocean and eastern Pacific Ocean.

Siliceous oozes are formed by the planktonic plants and animals consisting of diatom ooze and radiolarian ooze. Both the oozes have a predominance of silica content which does not dissolve in water. They are found at greater depths both in warm and cold waters. Radiolarian oozes is formed by shells of radiolaria which dries up in grey powder. It is mostly found between the depth of 2000 to 5000 fathoms in the tropical oceans. It predominates in the Pacific Ocean. Diatom ooze is formed by the marine plants containing silica in abundance. It is blue in colour but changes to yellow and coherent white powdery stuff when dried. Diatom ooze is found significantly around Antarctica, north Pacific from Alaska to Japan at a depth of 600 to 2000 fathoms.

(b) Red Clay

It is the most significant inorganic matter, a type of deposit which covers most of the deeper parts of the ocean beds. It is largely composed of iron oxides, aluminium silicate (85 percent) and contains volcanic material, especially pumice, ameteoric dust and occasionally the teeth of sharks etc. It is soft, greasy and plastic in character and due to the predominance of iron oxides it becomes uniformly red in colour. Red clay covers a very extensive area between 40 degree north to 40 degree south latitudes in the Pacific, Indian and Atlantic Ocean, covering about 1.29 million sq. km of area. It may be pointed out that red clay contains more radioactive substances than any other marine deposite. The marine deposit will be more clear from the chart given below.[62]

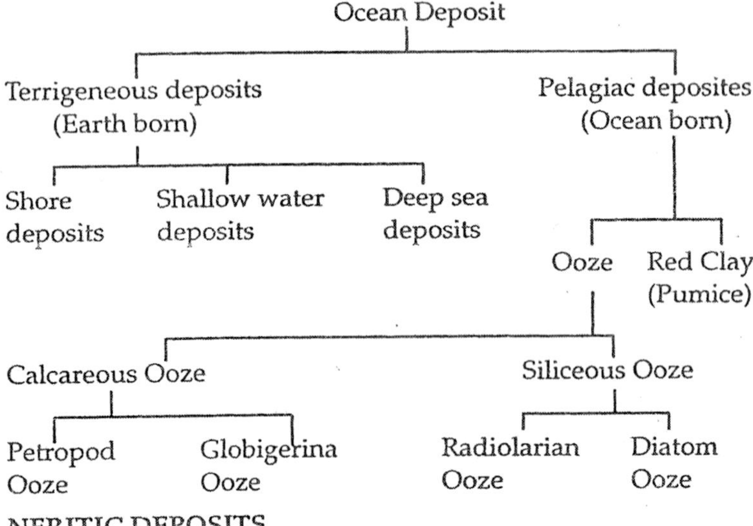

NERITIC DEPOSITS

Besides the two main marine deposits already discussed there is a third significant deposit known as neritic deposit. These deposits are produced by the orgnic shells and skeletons of various marine animals. The surface area of the oceans upto a depth of 100 fathoms is called Neritic province. Within this depth the sunlight is abundant which is necessary for the growth of marine animals. This zone has low salinity and has less turbulent motion. It is a store house of plant nutrients and diatoms, thus it become specially a natural home of fish. Deep water neritic deposits are found around the oceanic islands. Here, coral formation is of special mention.

CORALS

Coral is a small marine creature, having a hard skeleton, flourishes in shallow tropical seas. The skeleton is made of calcium carbonate extracted from the sea water in which the coral polyp lives. When the coral polyp dies, the softer parts of its body decay and are washed away, but the skeleton is left behind. According to M.S. Ladd and J.E. Hoffmeister (1936) the maximum depth for ideal growth of corals is about 100 to 150 fathoms while Gardinar gives a depth of 150 to 170 fathoms for the survival of coral polyps. The coral creatures require a sufficient amount of sunlight, oxygen and a mean annual temperature of 20 degree C for their survival. Hence they thrive in the tropical oceans confined between 25 degree south latitudes.

The muddy and turbid water clogs the mouth of corals and results into their death. They need clean and sediment free water but only lime is the important food for the coral polyps to live on. The warm oceanic currents supply necessary food favourable for the coral polyps, as they start the formation of their colonies from a firm base of had rock. They grow seaward side on submarine platform and beeches. When the coral polyps dies, the softer parts of the body decay and are washed away, but the skeleton is left behind. The increase in numbers of the coral polyps cause the masses of coral to grow to enormous size, forming coral Reefs and coral Islands.

CORAL REEFS

When the coral polyps die, their skeletons are deposited in layers and from hard coral rocks. The growth of these coral layers leads to the formation of a ridge known as coral reef. In other words, a chain of rocks lying at the surface of the sea and built up principally by immense numbers of coral skeletons. Both are piled up by wind and wave forming the new land. The coral formations are cemented by solution and calcareous sand. Coral reefs become masses of lunic stone and dolomite due to the pressure from above layers. These reefs receive deposits of sand and gravel on their upper portions which often come to be visible during the low tides as they develop more in width than in height. Sometimes these coral masses grow to enourmous size forming the coral reefs as coral islands.

CORAL ISLAND

A coral island is a coral reef situated far away from any other kind of land. It consist of a mound of sand resting on a flat coral reef several miles long and quite broad with a slight elevation. When there has been an uplift of the land, the coral island may be above sea level and may reach a considerable altitude. By various agencies like wind and waves, it may receive seeds, which become established and at length cover it with a rich vegetation.

Three more or less distinct kinds of coral reefs are recognized on the basis of nature, shape and mode of occurrence; (a) fringing reef, (b) barrier reef and (c) atolls.

(a) Fringing Reef

A reef formed by corals along the margins of continents and islands is called a frining reef. Its surface comprises a rough, uneven platform. Its seaward slope is steep and vertical while the landward gradient is gentle. Fringing reef is always attached to the coastal land but sometimes there is a shallow channel or lagoon between the platform and the land. Such lagoon is called 'boat channel', which is generally long but narrow in width. Its continuity is broken wherever the rivers drain into the sea.

(b) Barrier reef

An extensive coral reef that grows some distance away from coast or mainland is known as a barrier reef. A broad and deep lagoon separates it from the mainland. It is the largest, most extensive, highest and widest of all the three types of reefs and runs parallel to the coast for a long distance. The Great Barrier Reef parallel to east coast of Australia is the largest barrier reef of the world, which runs for a distance of 1920 km along Queensland.

(c) Atolls

An atoll is a coral reef in the shape of a ring or horseshore, enclosing a lagoon. According to Darwin's theory, an atoll began as a fringing reef round an island, then the island became submerged owing to subsidence, leaving only the ring shaped reef enclosing a lagoon. According to Murray's theory, an atoll was formed on the top of a plateau or hill which rose from the ocean bed to a depth at which the reef building corals live. The outer corals of a number of colonies grew most readily and reached the surface first, thus forming the atoll enclosing a lagoon. The atolls are found in Antilles Sea, Red Sea, China Sea, Australian Sea and Indonesian Sea. The atoll of Ellice Island (Funfuyi) is a famous atoll along which a lagoon of 12.8 km wide and 19.2 km long is enclosed.[63]

16
Atmospheric Pollution

DEFINITION AND CAUSES

The word 'atmospheric pollution', means uncleanliness vitiation and defilement of atmosphere around us, to which we can also name as environmental deterioration. One of the best definitions of pollution is given in the President's Science Advisory Committee (PSAC) 1965 report on "Restoring the Quality of our Environment", which is as follows.

"Environmental pollution is the unfavourable alteration of our surroundings, wholly or largely as a by-product of man's action through direct or indirect effects of changes in energy patterns, radiation levels, chemical and physical constitution and the aboundance of organisms".[66]

For an overall development of a country there is always a need to bring an industrial development. The industries raise per capita income and provide jobs to a large chunk of a unemployed youth of a nation. Thus industrial development is necessary for economic development. But this rapid industrialisation has brought in its wake the problem of large scale environmental pollution. The environment is in fact the biosphere i.e. a part of land, water and air where life exist. Any unwelcome changes within this space adversely affect the very existence of life. The biosphere can be adversely affected by these changes taking place for beyond its confined limits. For instance, the large scale operation of supersonic air craft flying through the ozone layer of atmosphere, reduces ozone concentration. Ozone shields the harmful untraviolet rays of sun reaching the earth, the decrease of which is therefore, a health hazard. Similarly dumping of radioactive wastes in deep oceans, in the long run, may have disastrous affects on marine life.

The increasing number need more food than the ecosystem can possibly cope with. The increasing use of chemical fertilizers, pesticides etc., though improve agricultural production, may cause polluted fruits, vegetables and local pollution of drinking water. Burning of fossil fuels to meet the energy needs and for increasing industrial output results in release of various combustion products into the atmosphere. The waste products include carbon dioxide, oxides of sulphur and carbon monoxides, which have adverse effects on climate. Thus it has created a great danger for human survival.

The industries puff-out smoke through their tall chimneys into the higher layers of atmosphere. This smoke adds carbon dioxide, different other gasses and solid particles to the atmosphere. It has disturbed the natural proportion of the gasses and dust particles of the air. It has been estimated that from 1860 A.D. till date carbon dioxide contents in the atmosphere has increased by 10 percent. If the rate of increase remains the same the temperature of the earth will rise to this extent that the ice of Greenland and Antarctica will melt. It will raise the level of the sea and will drown most of the coastal cities of the world. Hence, the industrial development creates ecological imbalance.

The motor-vehicles release poisonous gasses into the air which causes suffocation in breathing with certain dangerous chest diseases. It has been estimated that only in UK the industry in the country discharges 2 million tonnes of smoke, 1.5 million tonnes of ash and 5 million tonnes of sulphur with other gasses into the atmosphere every year.

Another smoke and toxic acids released by some industrial products is the refrigerator of industry, which uses the flourocarbon gasses. The release of this gas has caused thinning of ozone layer of atmosphere, which is so important in absorbing the harmful ultraviolet part of solar radiation reaching the earth's surface.

The industralisation pollutes the environment and disfigures the face of the nature, which is a positive danger to health. Workers living in a big industrial town live in slums and insanitary dwellings and in unhygienic conditions. Recently about 2500 people died of the poisonous gas leaking from Union Carbide Factory at Bhopal. More than a lakh were seriously injured besides thousands of cattle and livestock were killed. On the other hand, forests which purify the air have been recklessly cut down, causes soil erosion and floods. A recent study in national capital has revealed that for every death attributed to outdoor pollution, two fall victim of indoor pollution with women and infants being the worst effected. This pollution is mainly due to bio-fuels like dung cake, wood, charcoal and tyres of vehicles, which emit 23 tonnes of particulate matter per year.

TYPES

The purpose of this chapter is to highlight some of the important areas where ecology has relevance to human affairs. Hence, we will treat the different types of pollution i.e. water, air and noise pollution one by one separately as follows.

Air Pollution

Today in many parts of the world air pollution represents more serious threat to human existence. Major cities as well as rural areas of the world have serious air quality problems as their economy begins to prosper and expand. Air pollution arises from rapidly increasing automobile, truck, bus traffic and burgeoning industrial development. The capital of Nepal (Kathmandu) in the Himalayan mountains north of India, began to show symptoms of air pollution in 1971. The densely populated valley is often hazy with exhaust smoke from rapidly increasing automobiles and air traffic. Unless control measures are taken, the magnificent scenery which brings tourists to Nepal, will be viewed through a polluted haze in future years.

Another such example of air pollution is the famous London smog of 1952 in which about 5000 people died from respiratory distress in a persistent smog. And the crisis in Donora, Pennsylvania in 1948, in which hundreds of people suffered similar fatalities during and after a severe smog conditions for weeks together.

The clues to the health hazards of air pollution first came from clinical observations of increasing rates of emphysema, chronic bronchitis and respiratory distress in city dwellers. Sulfur dioxide is one of the common air pollutants which is most injurious to human health and impairs normal breathing. Dyspnea, bronchitis, cough, sputum production, wheezes, eye irritations and general malaize are the common diseases caused by air pollution. In an industrial complex of New Jersey, the death rate pollution (Medical World News 1967). In fact, respiratory ailments related to air pollution e.g. emphysema, chronic bronchitis, lung cancer and severe asthma are common in industrialized nations.

It has been estimated tht 164 million metric tonnes of pollutants enter the United States air every year (Newell 1971). Air pollution also takes its toll of buildings and other man-made objects. When moisure accumulates in polluted air, the oxides of sulfur, carbon and nitrogen form acids which are corrosive to metals, stone, paint, rubber and textiles etc. Fruits, vegetables and even deciduous trees have all fallen victims to air pollution in various parts of the world. Thus, there is no doubt that air pollution is a very significant and increasing factor in environmental deterioration. But certainly, many types of air pollution can be controlled by modern technology, besides more careful monitoring of automobile exhaust control system.

Water Pollution

It is a vast and generic term with a variety of meanings. It means any type of contamination between two extremes (1) a highly enriched biotic community with nitrates and phosphates (river and lake); or (2) a biotic community with sufficient concentration of toxic substances to eliminate living organisms. The types of water pollution may also be identified as surface water, underground water, marine or river water pollution. Excessive nutrients commonly originate in domestic sewage and run-off as agricultural fertilizer. Toxic chemicals originate in industrial operations, acid waters from mine seepage, surface erosion and washings of herbicides and insecticides.

Bacterial and viral contamination is a threat for the spread of water-borne diseases such as typhoid, dysentery, cholera etc. The Hudson river receives over 200 million gallons of sewage (PSAC Report 1965). Water-borne disease caused 20,000 illnesses and several deaths in California in 1965. The plants inactivate bacteria from water but remains still rich in basic nutrients. In 1962 about 38 states reported water pollution caused fish kills totalling 381 separate incidents.

Another major problem in water pollution is the addition of various ions and chemicals in the water which have toxic effects on plants, animals and human life. Chlorine control bacteria but cause mortality of fish in streams. In Japan, human illness and death occurred in 1950 among fishermen who ingested fish contaminated with mercury compounds from coastal industries. Like mercury there are so many other chemicals in water, polluted by industrial wastes which are toxic or pathogenic for man and animals. As a matter of fact more than 12000 toxic industrial chemicals are in use today.

A final facet of water pollution is the marine pollution. Though the oceans are virtually unlimited sources to accommodate the waste products of terrestrial man, there is substantial evidence of global pollution of coastal waters and open oceans. Mr. Jacques and Heyerdahl, the two famous explorers reported in 1971 that the oceanic waters are seriously polluted. They found waste oil and floating debris in the middle of the central Atlantic Ocean. Cousteau reported a 40 percent decline in marine life in 50 years of his experience on the world's oceans. He observed that coral reefs are shrinking over the entire world and more than 1000 species of sea life have become extinct in the last 50 years. Interaction between the sea and the atmosphere plays an important role in dispersing the pollutants in the sea. The oceanic currents, waves and winds over the seas are directly related to the accumulation of pollutants from the land to the sea. Air-borne particulate matter and gaseous pollutants can also find their way to oceans particularly when they are released into the coastal atmosphere.[67] Hence, from the above mentioned facts it is clear that modern man has contaminated the far reaches of the earth and thus the ecosystem of the world is reeling under the impact of pollution.

Seriously Polluted Rivers

We depend on running water for many things e.g. for recreation, energy, fresh water, and waste disposal. Unfortunately, our use of the world's streams and rivers has generally resulted in deterioration in the quality and quantity of water. Many rivers are no more than sewers and polluted rivers flowing into the sea having an increasingly detrimental effect on the ecological balance of marine life. Dams have been constructed across rivers without proper long-range planning as to their effect on plant and animal life, and even as to their benefits for people. The Aswan Dam built across the Nile River in Egypt in 1960, is a typical example.

More than fifty percent of the world's major rivers are seriously depleted and polluted, poisoning surrounding ecosystems and threatening the health of tens of millions of people. River systems are found in the worst state on both rich and poor countries. In America, the worst effected rivers include Colorado, which flows mainly in the S. Wn. United States and touches N. Wn. Mexico and the Lerma river, which flows through the middle of Mexico into lake Chapala. In Eurasia, the sickest rivers include AMU river and Syr river, which flow into the Aral sea in central Asia, the Ganges, which flows from the Himalayas to the Bay of Bengal and the Yellow river (severely polluted) which has its course through China, the most important agricultural region. All these rivers have been declared as 'very unhealthy' by the World Commission on Water (WCW).

The only major river systems which can be labelled 'healthy' are the Amazon in S. America and the Congo near Sahara (Africa). Both these rivers remain relatively untouched by industrial activity which has polluted other great rivers of the world. The Mekong river of S.E. Asia and the St. Lawrence river, which flow along the US-Canadian border are also better off than most of the other river systems around the world according to WCW.

Ismail Seraj-ud-Din, Chairman WCW says, "Overuse and misuse of land and water resources in river basins in both advanced industrial countries and developing countries constitute the primary cause for water pollution." According to the commission the river systems which have suffered the greatest damage are those whose water has been used to support intensive irrigation and industrialisation. The construction of multiple dams on many of these rivers has compounded the situation. This is true of yellow river, two years ago which ran dry its lower reaches for about 226 days in one year. In order to compensate for the lack of river water, the farmers were forced to draw up large amounts of water from aquifers, depleting them to increasingly dangerous levels. It all results in health problems of the region, which are growing day by day. The metals enter in the grain crops, vegetables and strawberries grown from polluted irrigated fields.[71]

Noise Pollution

In recent years, as serious as air and water pollution is to the health and welfare of man, the noise pollution represents an even more serious threat to human life. Excessive noise and unwanted sound has been a part of the industrial environment for a long time. The public is subjected to increasing noise from traffic, airplanes and electronic entertainments.

The sound is measured in terms of decibels, one decibel being approximately equal to the threshold of hearing in man. In a typical urban environment, background noise in a quiet sound protected room runs 40 decibels. On the other hand, ordinary street noises average 70 to 80 decibels. Around the home, background noise averages 40 to 50 decibles. The general conversation produces 60 decibels, a garbage disposal 85, a vacum cleaner 90 and a power lawn mover 105 decibels. Heavy city traffic produces 95 to 110 and a jet aircraft taking off generates 120 to 150 decibels. In India, the capital city (Delhi) with 83, Calcutta 82, Mumbai 80 and Chennai 77 decibels of sound has also fallen under the menance of noise pollution (IRTE).

According to the recent researches of medical sciences, noise can be a significant nervous stress, as it is an irritant and source of environmental annoyance. It can increase irritability and reduce job efficiency. Some cases of noise pollution has shown changes in heart rate, blood pressure and metabolism similar to other types of emotional anxiety and stress to man. The hearing loss and even early deafness can be caused by prolonged noise about 95 decibels. This is why, the workers in noisy factories are often subject to hearing loss.

The noises in a big city particularly are deafening and unbearable. The rush and roar of vehicular traffic, the loudspeaker and the beat of drums are all very annoying. The sick and old, the students and research scholars cannot apply their minds to anything when they are surrounded on all sides by disturbing sound and noises. The modernity or development continue generating huge levels of decibels round the clock. They have reached dangerous proportions as to have begun to prove a health hazard to the common man.

What happens when a person is exposed for a long to noise levels above 50 decibels ? Doctors say, finer hearing abilities not only get impaired for good but even a person's work efficiencey goes down. He falls victim to a host of ailments like fatigue, insomnia, mental depression, digestive disorders, hypertension and cardiac problems.

The bursting ear-splitting crackers at any function, merrily hanking and hooting sound of horns by drivers, cinema dialogues, songs the televisions and the radios at top volume (even in buses) all create noise pollution everywhere. This sound and noise of high intensity is very dangerous and harmful especially for children, pregnant women, sick and elderly people. With the air and rail travel growing more popular, the people living near airports and busy railway tracks have been found to suffer from disorders of the nervous system.

Remedies of Pollution

The widespread and increasing pollution must be checked in time to save the mankind. In order to check the pollution, there should be balanced and planned industrialisation. Urbanisation should also the well-planned and diversified. The industrial waste should not be dumped on the earth or in the rivers. It should be suitably burnt or buried deep into the earth. An elaborate programme of planting trees and plants should be taken in hand. Deforestation should be strictly prohibited under law. Polluted water should not be used for irrigation purposes. People should be made aware to feel their moral responsibility to keep the atmosphere neat and clean.

In western countries, blowing of horns is banned and that is the only way to end such menace of noise pollution that goes with it. The permanent out of door loud speakers on public place should also be banned by law. They should only be allowed to fix them inside their interiors for the benefit of listeners. The workers in noise-generating factories should be provided with ear-muffs or ear-plugs to cut out the harmful noise. The loud crackers and fire works should be totally banned as they have both, high contents of SO_2 and CO_2 and high levels of noise (more than 100 decibels) against the normal tolerable limit of 50 db. If this all is not checked and controlled in time the earth will be reduced to a barren land as the pollution poses a big threat to the world civilization and culture.

Some of the ways and means for checking and controlling the menace of air, water and ground pollution as a result of the growth of industrial culture are as follows.

(a) Installation of pollution control devices in the polluting industries.

(b) Adoption of latest technology by the Industrial Unit holder.

(c) To control the air pollution due to smoke of vehicles, the Prozone Pollution Device Manufacturing Units should be set up in Industrial areas. It can also generate employment to the youth.

These devices patented and manufactured in the United Kingdom and the United States of America, are stated to curtail fuel consumption by 15 percent in the automobile vehicles and simultaneously reducing the emission of Carbon Monoxide, Lead Oxide etc. to a considerable level. The test certificates and report of ARAI, Pune (Maharashtra), indicate that the devices are highly effective both in reducing fuel consumption and controlling emission of the pollution.[72]

Glossary

Nebula
A large cloud of dust and gas among the stars, originally the term was applied to any hazy celestial object, including huge, distant systems of stars now called galaxies.

Fault Plane
Refers to the surface along which a fault has occured.

Fault Scarp
A scarp situated where a fault has occured. It is due to the relative downward movement of the strata on the lower side of the fault.

Earthquakes
A sudden and violent shaking of the earth as a whole or a part of it, caused by volcanic erruption, folding, faulting and slipping of the crust.

NE 'VE' FRENCH FIRM GERMAN

The granular substance, partly snow and partly ice, which is formed as the snow that has accumulated at the head of a glacier valley is being transformed into glacier ice.

Snow-Guage
An instument used for measuring the depth of snowfall.

Avalanche
A vast mass of snow and ice at high altitude which has accumulated to such an extent that its own weight causes it to slide rapidly down the mountain slope, often carrying with it thousands of tonnes of rocks.

Arete French

A sharp mountain rigde, often formed by the erosion of two adjoining cirques.

Cirques or Corrie

A deep, rounded hollow with steep sides, formed with erosion by snow and ice in a glaciated region. Often the cirques are filled with snow water as lakes which enhance the beauty of the mountain scenery.

Debris or Detritus

Refers to a collection of material formed by the disintegration or wearing away of rocks.

Grazing Fields

The vast natural grass lands or grounds on the mountain slopes of Himalayas and Alps, used for cattle grazing and rearing. They stretch over hundreds of miles in length and breadth.

Magma

The molten material which exists below the solid crust, often comes out on the surface of the earth through volcanic explosion and dries in volcanic rock.

Karst Region

A limestone region in which all the drainage is by underground channels. Thus, the surface being dry and barren, named after karst district of Alps, near Yugoslavia.

CRATER

The funnel shaped depression on the top of volcanic mountain. The bottom of the funnel opens into the funnel or pipe through which the erupted material finds its way to the surface. An extinct crater is often filled with snow or rain water farming a lake on the mountain top. Kailash lake in Bhadarwah (J and K) is the typical example of such lake.

Crest and Trough

Formed by the rise and fall of the surface water of the ocean, when winds blow over it. The highest point of water is known as crest while the lowest point as trough. This swinging of surface water to and fro is called as wave.

Breakers

Refers to a wave that breaks into foam when it advances towards the shore, where the water becomes so shallow that it is insufficient to complete the wave form.

Notch

Refers to the undercutting of a cliff at high water mark may produce a notch at its foot.

Nip

Refers to a distinct break of slope at the higher edge of a beach, related to high water mark, just a few feet above the level of the highest tides.

Lagoon

A shallow stretch of water which is partly or completely separated from the sea by a narrow strip of land.

Compressive Forces

If the movement of the earth's crust is towards a common point from two opposite sides, the force is known as convergent or compressional force. It results in crustal bending or folding of mountains.

Convection Current

The transmission of heat from one part of liquid or gas to another by the movement of particals themselves. When its lower portion is heated, it expands and rises up replaced by cool mass of air or fluid. The air is heated by contact with earth's surface and rises up as convection current.

Abyssal Plain

The term means the plain regions in the lowest depths of oceans, with an average depth of 2000 fathoms. These marine plains are extremely vast with no plant life.

Fathomometer

An instrument used to determine the depth of the ocean, making use of the known velocity of sound waves through water. One fathom is equal to six feet.

Typhoon

Refers to a small, intense, vertical tropical storm in the China sea and along the margins of the west Pacific ocean, accompanied by winds of terrific force (above 100 miles per hour), torrential rain and thunderstorm.

Meteorology

The science which investigates the atmosphere and the physical processes which occur therein. The subject therefore, includes the study of atmospheric pressure, temperature, winds, precipitation, cloudiness and sunshine etc..

Frost

The particles of frozen moisture formed on the earth's surface on the blades of grass or other objects.

Hail

The hard pellets of ice with a diameter of about 5mm which fall from cumulonimbus clouds, often associated with thunderstorms. They are often caused by rapid ascent of moist air in the atmosphere.

Fog and Mist

A dense mass of small water drops or smoke or dust particles in lower layers of atmosphere. It is termed as fog when the visibility is upto 1 km and is termed as mist if the visibility is less than 1 km.

Dew

The moisture deposited on the blades of grass or other objects on the earth's surface, when nocturnal radiation from the earth has cooled the lower layers of atmosphere below the dew point and the water vapour present has condensed into drops.

Snow and Ice

Precipitation which takes the form of white, delicate feathery ice crystals, when the temperature in the atmosphere is below freezing point. It accumulates on the surface of the earth as snow falls and the lower layers melt due to the above pressure and are transformed into ice.

Rain

Separate water drops which fall to the earth from the clouds, having been formed by the condensation of water vapour in the atmosphere.

Marine or equable Climate
A climate which is mainly influenced by the proximity of the sea. The sea is warmed and cooled less readily than the land and its influence tends to create equable climate.

Continental Climate
A type of climate experienced in the interior of greate continents. It is characterized by extremes of temperature, with maximum in summer and minimum in winter. Besides the great variation in both diurnal and seasonal range of temperature is the chief characteristic of continental climate.

Monsoon
A term derived from the Arabic word 'Mawsim', meaning season, applied by the Arabs to the seasonal winds of Arabian Sea, which blow for about six months from the southwest and six months from the northeast.

Relative Humidity
The ratio between the actual amount of water vapour in the given volume of the air and the amount which would be present if the air were saturated at the same temperature, usually expressed as a percentage. It is determined by means of an instrument known as hygrometer.

Condensation
The process by which a substance changes from the vapour to the liquid state. The opposite process to condensation is termed as evaporation by which clouds are formed.

Orbit
The oval path through space on which the planets revolve round the sun.

Solar Constant
The intensity of the sun's radiation in space at the mean distance of the earth from the sun, usually expressed in calories per sq cm per minute.

Lapse Rate

The rate of change of temperature with height in the atmosphere, being usually expressed in F degree per thousand feet or C degree per hundred meters.

Range of Temperature

The difference between highest and lowest temperatures of a place during a certain period. The difference between the mean temperatures of the warmest and coldest months is known as mean annual range of temperature.

Isotherm

A line on the map joining the places of the same temperature at a particular instant or over a certain period.

Inversion of Temperatures

An increase of temperature with height above the earth's surface, being the reverse of the normal situation, in which the temperature falls with height.

Knots

A unit of speed equal to one Nautical mile per hour.

Anemometer

An instrument by which the velocity and direction of the wind is measured, usually in miles per hour or meters per second.

Temperate Zone

A zone in the middle latitudes between Torrid and Friged zones. In other words a zone between Tropic of Cancer and Arctic Circle, in northern hemisphere and between Topic of Capricorn and Antactic Circle. The sun is never overhead in this zone.

Torrid Zone

The region lying between the tropic of cancer and tropic of Capricorn is known as torrid zone. Within this zone the sun is directly overhead twice a year, its rays are never very oblique and the weather in general is always hot. It includes adjacent areas of rather higher latitudes i.e. 30 degree north and 30 degree south.

Tropic of Cancer

It refers to the parallel of latitude roughly 23½ degree north, showing the extreme northern positions at which the sun seems directly overhead at noon. The sun's rays shine vertically on the tropic of cancer at the summer Solstice of northern hemisphere.

Tropic of Capricorn

It refers to the parallel of latitude, roughly 23½ degree south, indicating the extreme southern positions where the sun seems directly overhead at noon. The sun's rays shine vertically on the tropic of Capricorn at the winter Solstice of the southern hemisphere.

Prevailing Winds

It refers to the wind, shown by direction at a certain area that has a considerably higher frequency than any other.

Doldrums

The equatorial belt of low atmospheric pressure where the northeast and southeast trade winds converge on and meet each other, producing calms and light surface winds and a strong upward movement of air, known as doldrums. These winds are characterized by turbulent and stormy weather, with heavy rains, thunder storms and squalls.

Land Breezes

A diurnal wind blowing from the land out to sea. It is caused by the differencial cooling of land and sea. The cooling of the air over the land by radiation during the night causes the air to descend and flow seawards. It is much influenced by topography.

Sea Breeze

The diurnal movement of air from the sea to the land, caused by differential heating. During the day, the greater heating of land causes the air to ascend and the air from the sea moves in to take its place.

Occlusion

In a depression, the closing of the cold front on to the warm front. The warm sector of the depression is lifted by the advancing cold air and the earth's surface is reduced to a line (occlusion). It exists in the upper air, however, the cloud and rain of the original fronts remain with it for some time.

Horizon

A great circle on the celestial sphere, parallel to the visible horizon of an observer. In other words a circular line bounding an observer's view, where earth and sky appears to meet. Its distance is determined by the height of the observer.

Icebergs

When a glacier enters the sea, a mass of ice is broken off and floats away on the surface of the ocean water is known as iceberg.

Nautical or Geographical Mile

A unit of distance used in navigation, equal to the length of one minute (1′) of arc of a Great Circle drawn on the sphere having the same area as the earth. This is approximately 6080 feet, while the statute mile is equal to 5280 feet (1760 yards).

Meteor or Shooting Star

A fragment of solid mater which enters the upper atmosphere from outer space and becomes visible through incandescene caused by the resistance of the air to its passage. It travels upto 50 miles per second and usually becomes luminous during its descent. Mostly they disintegrate during their passage and reach on the surface as dust and ash, except a large body known as a meteorite, which falls solid to the ground.

Fathom

The unit of length, equal to 6 feet, used mainly in determining the depth of the sea.

Lagoon

A shallow stretch of water which is partly or completely separated from the sea by a narrow strip of land. In the case of coral reef, it is a channel of sea water between the reef and main land.

Decidious Forest

The trees which shed their leaves in one season (autumn) of the year, while on the other hand evergreen trees do not shed the leaves at once.

Hurricane

A very forceful wind having a mean velocity of over 75 miles per minute. These are the tropical cyclones of West Indies, Gulf of Maxico, coast of Queenland and Australia.

Typhoons

The tropical cyclone of the China Seas is called typhoon. It is experienced most frequently during the late summer and early autumn. The Philippine Islands and coastal areas of southern China lie in the direct track of many of the typhoons. They bring winds of tremendous strength, torrential rain and destruction.

Westerlies

The westerly winds which blow with great frequency in regions lying on the poleward sides of the sub-tropical high pressure areas (Horse Latitudes). In winter they move southwards effecting Mediterranean regions while in summer they move northward in northern hemisphere.

IRTE

Institute of Road Traffic Education.

Dykes and Sills

A vertical sheet of igneous rock, formed by moulten rock or magma. When the magma is forced from the interior towards the surface of the earth, it cools and solidifies in its passage and becomes dyke rock. The dyke may change its direction or branch away into smaller dykes or give rise to sills. Its thickness varies from an inch to hundreds of feet.

Volcano

A vent in the earth's crust caused by magma forcing its way to the surface. The lava is finally ejected with explosive force, rock fragments and ashes being thrown into the air. The emissions of lava often cause the volcano to take the form of a conical hill or mountain.

Volcanic Neck or Plug

A mass of solidified lava filling the central vent of a volcano. It is sometimes left isolated after the remainder of the cone has been removed by denudation.

Volcanic Cone

A cone formed on the top of a volcano by the accumulation of errupted material.

Volcanic Ash and Dust

Fine particals of lava ejected from a volcano in eruption are known as ash, while these are called as dust when they are blown into the air.

Caldera

A large basin-shaped crater bounded by steep cliffs, usually formed by the subsidence of the top of a volcanic mountain and sometimes occupied by a lake.

Geyser

A hot spring which at regular intervals throws a jet of hot water and steam into the air. Sometimes the jet rises to a height of 100 to 200 feet. It usually occurs in a volcanic region.

Hot Springs

Hot springs are not always confined to volcanic regions as water sinking far enough to the earth may become heated by chemical reaction and rise to form springs. Various skin diseases are cured by hot spring water.

Knoll

A knoll is an isolated or detatched hill having a marked low height.

Cirque

It is also called Corie or Kar. In a glaciated mountainous region the old catchment basins of glaciers have the appearance of steep wall arm chair hollows which are called cirques. It generally has a lake at its centre with a well developed rim all around it. The lake is called "Tarn".

Statistics of the Earth

Diameter

The mean (average) diameter of the earth is 7,913 miles. The diameter at the equator is 7926.42 miles and at the poles is 7,899.83 miles.

Circumference

The distance around the equator is 24,902 miles and around the poles it is 24,860 miles.

Surface Area

The area of the earth's surface is about 197,000,000 square miles, of which 58,000,000 sq miles are land and 139,000,000 sq miles are water.

Density

The average density of the earth is 5.522 times that of water, or 344.7 pounds to the cubic foot.

Weight

The total weight of the earth is estimated at 6.586×10^{21} tons or 6,586 billion billion (18 zeros) tons.

Distance from the Sun

The average distance of the earth from the sun is 92,9000,000 miles.[73]

Some Useful Imperial and Metric Equivalents & Conversion Factors

(a) Weight

1 ounce	28.350 grams = 437.5 grains
1 pound	0.4536 kilogram = 16 ounces
1 short cut	100 pounds = 45.359 kilogram
1 long cut	112 pounds = 50.802 kilogram
1 short ton	2000 pounds = 0.9072 tons
1 long ton	2240 pounds = 1.0161 tons
1 ton	1016 kilograms
1 metric ton	1000 kg = 1.1023 short tons = 0.9842 long tons
1 short ton	2000 lbs = 0.9072 metric tons
1 long ton	2240 lbs = 1.0161 metric tons
1 hectare	2.4711 acres
1 sq km	100 hectares = 247.1054 acres = 03861 sq mile
1 acre	0.4047 hectares
1 sq mile	640 acres = 2.5900 sq km = 258.9988 hectates
1 kilogram	2.2046 pounds

(b) Linear measure

1 inch	2.54 centimetres
1 foot	0.3048 metre
1 yard	0.9144 metre
1 mile	1.6093 kilometers
1 mm	0.0394 inches
1 centimeter	0.3937 inch = 10 mm
1 metre	3.2808 feet = 1000 mm 1.0936 yards
1 kilometer	0.6214 mile = 1000 meters

(c) Capacity

1 cubic centimeter	=	0.0610 cubic inch
1 cubic decimeter	=	1000 cubic centimeter
1 cubic metre	=	1000 cubic decimeter
1 litre	=	1 cubic decimeter
	=	0.2642 U.S. Gallon
	=	0.2200 Imp. Gallon
1 Hecto litre	=	100 liters
	=	2.8378 U.S. Bus
	=	2.7497 Imp. Gallon

(d) Area

1 Square	=	2.54 centimeter
1 foot	=	12 inch = 0.3048 centimeter
1 yard	=	1760 yards
1 Mile	=	1.6093 kilometers

References

1. O.P. Verma, Geography Teaching, pp. 3,4.
2. New Standard Encyclopedia, Vol. 6, pp. G-77-78.
3. New Standard Encyclopedia, Vol. 6, p. G-77.
4. M.S. Rao, Dictionary of Geography, p. 307.
5. Hartshorne Richard, The Nature of Geography Annuals of the Association of American Geographers, Vol. 29, No. 3 and 4. Lancaster Pennsylvania USA.
6. Welpton, W.P., "The Teaching of Geography,", University Tutorial Press Ltd., London, p. 3.
7. Smith Russell, Industrial and Commercial Geography", Constable and Coy Ltd., London, p. 3.
8. A. Das Gupta, A.N. Kapoor, Principles of Physical Geography, p. 1.
9. Dr. R.C. Bhagt, Dr. Raina, Principles of Geography, pp. 4-5.
10. New Knolwdge Library, Universal Reference Encyclopedia Vol. 28, p. 2620.
11. Savindra Singh, Physical Geography, p. 4.
12. New Knowledge Library, Vol. 8, p. L-79 b.
13. New Standard Encyclopedia, Vol. 11, p. 394.
14. George Gamow, 1962, Biography of the Earth (Macmillan, London), p. 32.
15. *Ibid.*, p. 32.
16. W.M. Smart, 1955, The Origin of the Earth, *op. cit.*, p. 204.
17. Stars and Planets, Macdonald Junior Reference Library, p. 53.
18. *Ibid.*, p. 53.
19. Peter Hariss, 1973, *op. cit.*, p. 56.

20. New Standard Encyclopedia, p. E-12.
21. Bullard, Everett and Smith, (Royal Society London) Symposium on Continental Drift; *op. cit.*, pp. 8, 41, 51.
22. Holmes, A. Principals of Geology, Nelson London, *op. cit.*, p. 1197.
23. Holmes, A, 1966, Principals of Physical Geology (Nelson London), p. 27.
24. A., Dass Gupta and A.N. Kapoor, Principles of Physical Geography, pp. 39, 40.
25. Savindra Singh, 1991, Environmental Geography, pp. 68.
26. M.S. Rao, Dictionary of Geography, p. 246.
27. W.S. Moore, Dictionary of Geography, p. 38.
28. A.K. Dutta, Introduction to Physical Geology, p. 186.
29. Jasper H. Stembridge, the World, a General Regional Geog., p. 20.
30. A.K. Data, Introduction to Physical Geology, pp. 162-64.
31. *Ibid.*
32. Sailender Raghav, Refresher Course in Geog, pp. 61-63.
33. Prece and Wood, Foundtion of Geog, pp. 75-76.
34. Soh Cheug Leong, Certificate Physical and Human Geography, p. 13.
35. Shailendra Raghav, Refresher Course in Geog, p. 33.
36. Richard, S. Palm, Physical Geog, p. 250.
37. *Ibid.*; p. 245.
38. Savindra Singh, Physical Geog, p. 60.
39. Savindra Singh, Physical Geog, p. 203.
40. Sailendra Raghav, Refresher Course in Geog, p. 113.
41. Savindra Singh, Physical Geog, 1973, p. 238.
42. *Ibid.*, p. 239.
43. Goh. Cheng Leong, Certificate Physical and Human Geography, p. 29.
44. W.G. Moore, A Dictionary of Geog, p. 163.
45. A.K. Datta, Introduction to Physical Geology, p. 89.
46. Abid, p. 122.

47. W.G. Moore, A Dictionary of Geog, p. 63.
48. Goh Cheng Leong, Certificate Physical and Human Geog. (Indi2an Edition), pp. 49-50.
49. Richard S. Palm, Physical Geog, p. 284.
50. A.N. Kappor, Principal of Geog, p. 255.
51. M.S. Rao, A Dictionary of Geog, p. 66.
52. R.S. Palm, Physical Geog, p. 290.
53. S. Singh, Physical Geog, pp. 273-74.
54. W.S. Moore, A Dictionary of Geog, p. 188.
55. Savindra Singh, Physical Geog, pp. 356-59.
56. W.S. Moore, A Dictionary of Geog, pp. 38-39.
57. Dr. R.C. Bhagat, Principles of Physical Geog, pp. 136-37.
58. M. S. Rao, Dictionary of Geography, p. 183.
59. Sailendra Raghav, Refresher Course in Geog, p. 182.
60. Dr. R.C. Bhagat Principles of Geog, p. 167-168.
61. New Standard Encyclopedia, Vol. 10, p. 18.
62. A.N. Kapoor, Principles of Geography, p. 295.
63. Savindra Singh, Physical Geography, p. 393.
64. George C.D. Long, Weather and Climate, p. 150.
65. Shaildendra Raghav, Refresher Course in Geog, p. 236.
66. Charles H. Southwick, Ecology and the Quality of Our Environment, p. 6.
67. A.A. Rama Sastry, Weather and Weather Fore-Casting, p. 165.
68. Jasper H. Stembridge, The world, A General Geography, p. 1.
69. New Standard Encyclopedia, Vol. 10, pp. 18-19.
70. S. Anand, Physical Geography and Geography of India, p. 161.
71. World Commission on Water, Washington, Dec. Ist, 1999, Kashmir Times.
72. Kashmir Times News Service, Jammu, Dec. 2nd, 1999.
73. New Standard Encyclopedia, Vol. 5, p. E-8.